WORKING *the* ROUGH STONE

WORKING *the* ROUGH STONE

Freemasonry and Society in
Eighteenth-Century Russia

Douglas Smith

NORTHERN ILLINOIS UNIVERSITY PRESS

DeKalb 1999

© 1999 by Northern Illinois University Press
Published by the Northern Illinois
University Press, DeKalb, Illinois   60115
Manufactured in the United States
using acid-free paper
All Rights Reserved
Design by Julia Fauci

Library of Congress
Cataloging-in-Publication Data
Smith, Douglas, 1962–
Working the rough stone : freemasonry and
society in eighteenth-century Russia /
Douglas Smith.
        p.   cm.
Includes bibliographical references
and index.
ISBN 0-87580-246-X (alk. paper)
1. Freemasonry—Russia—History—18th
century. 2. Russia—Intellectual life—18th
century. 3. Russia—Social conditions—
To 1801. 4. Enlightenment—Russia. I. Title.
HS624.S58   1999
98-32115
366'.1'094709033—dc21
CIP

TO MY MOTHER AND FATHER

Contents

Acknowledgments

A trip to the opera provided the initial inspiration for this book. It was at a performance of *The Magic Flute* at the Wiener Staatsoper in the winter of 1983 that I first became interested in the subject of Freemasonry. Although I had already read and studied Johann Immanuel Schikaneder's libretto as an expression of the Masonic movement—of its general character, ideals, and overall worldview—in one of my university courses, it was this performance that brought Freemasonry to life for me, suggesting in dramatic and powerful fashion the great importance these ideas possessed for the men, and women, of Mozart's age.

I would first like to thank those persons who were instrumental in getting me to the opera, so to speak. At Blake School, Rod Anderson and Heinz Otto introduced me to the study of history and foreign language and helped me begin to see the pleasure that such study can provide. At the University of Vermont, everyone in the Department of German and Russian created a uniquely stimulating and nurturing environment. In Vienna, Drs. Bernhardt and Strobele and, most especially, Anna Kogan taught me to appreciate Vindobona's charms from the perspective of the native and the alien.

More recently, I owe a debt of gratitude to John Brewer, John Hatch, Gail Lenhoff, Peter Reill, Geoffrey Symcox, and Ronald Vroon under whom I studied at UCLA. Margaret Jacob, Vicky Lettmann, Peter Pozefsky, Willard Sunderland, and Kevin Thomas all read versions of this book at various stages and provided useful comments and suggestions. Elise Wirtschafter was kind enough to read the manuscript twice and to offer needed criticism and support. The two reviewers for Northern Illinois University Press quite rightly pushed me to expand my topic and to develop further my arguments, helping to make for a better work. I am grateful to Mary Lincoln at Northern Illinois University Press for her conscientiousness and professionalism. I happily thank those friends and colleagues in both the United States and Russia who have so freely offered assistance in all manner of ways: Stephen L. Baehr, Barbara Bernstein, Michael Biggins, Aleksandr Bobosov, Mireille Dovat, Bob Edelman, James Goodale, Jason Horowitz, Igor and Alla Kotler, Aleksei Koval'chuk, Gary Marker, Tania Safronova, Andrei Serkov, Daniel Simhon, Nikita Sokolov, Oleg Solov'ev, David Spaeder, Theofanis Stavrou, Ol'ga Var'iash, Vera Wheeler, Glennys Young, Steve Zipperstein, and Andrei Zorin. My largest debt is to Hans Rogger, my graduate advisor, for all the help, advice, and encouragement he has given me over the past decade.

I acknowledge the assistance of the very considerate staffs at the numerous libraries and archives where I worked in Los Angeles, Moscow, St. Petersburg, and Seattle. Gerry Stigberg at the University of Illinois, Urbana-Champaign, was most helpful in acquiring illustrations for the book. I also acknowledge the generous financial support of the UCLA Center for Russian and East European Studies, the UCLA Department of History, M·O·M, the Kennan Institute for Advanced Russian Studies, the Fulbright-Hays Doctoral Dissertation Research Abroad Fellowship, and the International Research and Exchanges Board (IREX), with funds provided by the National Endowment for the Humanities, the U.S. Information Agency, and the U.S. Department of State, which administers the Russian, Eurasian, and East European Research Program (Title VIII).

Finally, no one has had to live more closely with this work—and its author—than Stephanie Ellis Smith, my wife. Thanks again, Stephanie, you make it all possible and worthwhile.

A NOTE ON DATES AND SPELLING

Until 1918, Russia used the Old Style (Julian) calendar. In the eighteenth century this calendar was eleven days—and in the nineteenth, twelve days—behind the New Style (Gregorian) calendar used in the West. All dates in this study are given in the Julian, or Old Style, calendar. Transliteration of Russian is by the Library of Congress system, although the names of certain figures and places have been anglicized. The original spelling of words and titles has generally been followed throughout the text and bibliography.

WORKING *the* ROUGH STONE

Ernst: Nun was ist sie denn, diese notwendige, diese unentbehrliche
         Freimäurerei?

Falk: Wie ich dir schon zu verstehen gegeben: —
         Etwas, das selbst die, die es wissen, nicht sagen können.

Ernst: Well, what is it then, this essential, indispensable Freemasonry?

Falk: As I have already intimated to you—
         Something that even those who know it cannot say.

— Gotthold Ephraim Lessing, *Ernst und Falk: Gespräche für Freimäurer*

Introduction

On 23 April 1792, Prince A. A. Prozorovskii, the hated governor-general of Moscow, dispatched a squadron of twelve hussars under the command of Major Prince Zhevakhov to arrest the feared Freemason Nikolai Novikov. Zhevakhov and his men rode all night before reaching Avdot'ino-Tikhvinskoe, Novikov's local estate, where they found Novikov sick in bed, slipping in and out of consciousness. Despite his poor health, Novikov was seated in a covered wagon and returned to Moscow for questioning by Prozorovskii. Within days Empress Catherine II (the Great) ordered Novikov conducted under heavy guard to St. Petersburg for interrogation by the notorious inquisitor S. I. Sheshkovskii. For the next several weeks, Novikov was questioned about his activities as a leading member of the Freemasons and about the connections of Russian Freemasons with their colleagues at the hostile court in Berlin and with the Grand Duke Paul, Catherine's estranged son and heir to the Russian throne. Novikov eventually confessed his guilt and admitted that he had tried to lure the Grand Duke into the Masonic order and that he had published banned books. In August, Catherine sentenced him to fifteen years imprisonment in the Schlüsselburg fortress. With the accession of Paul in 1796, Novikov was pardoned and released. He left prison a broken man, turning his back on society for his beloved estate where he lived out his final years in quiet seclusion.[1]

Publisher, journalist, historian, and Freemason, Nikolai Novikov was one of the most dynamic figures in the Russia of Catherine the Great (1762–1796). Generations of Russians have revered him as a veritable saint, and he is still referred to as an "Apostle of Good" and the "Enlightener of Russia."[2] His tragic story has long been synonymous with the history of Russian Freemasonry not merely because he was one of the most visible Masons, but more significantly because his arrest coincided with the demise of the Masonic movement. Not long after Novikov disappeared behind the walls of the Schlüsselburg fortress, Russia's few remaining lodges closed their doors, and what had once been a vibrant Masonic community vanished.

The history of this community represents a fascinating and important chapter in Russia's past. As the country's first widespread social movement, Freemasonry exercised a profound influence over Russian society.[3] Thousands of Russians joined the order over the course of the eighteenth century. Members ranged from rulers and mighty aristocrats to humble craftsmen and emancipated serfs. Freemasonry touched many of the age's leading statesmen, officers, thinkers, writers, artists, and merchants, who together created an extensive network of lodges that eventually spanned almost the entire empire. A look at these lodges will provide access to some of the most mysterious territory of eighteenth-century Russia. By observing the brothers at work one learns what it meant to be a Freemason and gains a sense of the desires and fears that motivated these men. Moreover, given Freemasonry's great popularity and recognized influence, to see the world as the

The "enlightener of Russia," N. I. Novikov (1744–1818) possesses an almost mythic aura due in large part to his arrest in 1792. His commitment to publishing, journalism, and education was matched by an equally strong dedication to Freemasonry that brought Novikov to the order's highest levels.

Masons saw it provides insight into matters of broader significance for the history of imperial Russia (1682–1917), such as the still dimly understood process of Westernization, the dynamics of group formation and social identity, and the nature of society.

Introduced by foreign travelers in the early decades of the century, Freemasonry was one of many imports to wash up on Russia's shores in the wake of Peter the Great's revolutionary reign (1682–1725). Intent upon remaking medieval Muscovy in the image of modern Europe, Peter initiated a crash course of Westernization that replaced traditional habits and customs with European military, political, economic, and social models. Peter's wholesale reforms had their greatest impact on Russia's elite, who were expected to abandon the ways of their ancestors and adopt the habits of the educated Dutchman or Swede. This meant more than shedding one's caftan and shaving one's beard; it meant learning to think and act in a new fashion. As the century progressed, Russia's educated classes came to share in the cosmopolitan culture of the European elite. They read, studied, and traveled; they acquired a different set of manners and new ideas about behavior itself; they visited salons and attended plays and concerts. And they became Freemasons.

Russians joined the order for the same reasons their European brothers did. Freemasonry provided a place to socialize with like-minded men, to cultivate patrons, and to enjoy a meal and a drink (or two) free from the burdens of work and family. It was also a place to imbibe more intoxicating spirits, be they of the Enlightenment, Christianity, or even the occult. But Russians also joined for reasons that, although not unknown to their European brothers, were different and especially important. In the lodges Russians acquired a new, Western standard of behavior. They learned to become civil and polite by curbing their base desires and passions, a process they called "working the rough stone." Along with such lessons, Russians received instruction on civic responsibility. Freemasonry taught the importance of serving state and society and helped the brothers to become dutiful officials, loyal subjects, and good citizens as well as caring spouses, kind fathers, and true Christians. In short, Freemasonry helped its members improve themselves and reassured them that they were indeed better men than their fellow Russians.

The Masonic message struck a chord in Russia where personal and group identity were ill-defined. Russia lacked the relatively clear boundaries that delineated social groups in western European countries. Rather, Russian society exhibited an atomized and fractured character that made social identity indeterminate and problematic.[4] Peter's reforms exacerbated this situation in a number of ways. Westernization meant the imposition of an entirely new system for marking social boundaries and ascribing social status. Education, dress, manners, and morals became the primary components for constructing identity and the chief criteria for assessing the worth of others. Peter further complicated matters in 1722 when he created the Table of Ranks, a system that replaced birth with state service as the foundation for social standing. To modernize Russia, Peter needed a class of diligent, educated men who would become the officials of the rationalized, bureaucratic state apparatus he was building. Established to further the common good by exploiting the country's untapped resources, Peter's "well-ordered police state" would reward its most dedicated officials with ranks and wealth in return for their service.[5] Noble ancestry was no longer enough to guarantee one a privileged place, and over the century nobles increasingly had to compete with educated commoners for ranks and titles, and the wealth and prestige these positions conferred. Even though Russia's nobility fought to be free of obligatory state service (a right gained in 1762), the ethic of service remained central to its self-definition. All of these factors— Russia's porous social boundaries, the complexities of Westernization, and the imperative of service—combined to make social identity a protracted and pervasive problem throughout the eighteenth and nineteenth centuries.

Many Russians found a haven from these pressures in Freemasonry, which provided them a bond of fellowship within a world of perceived chaos and disorder. Freemasonry gave them a sense of community, a feeling of being part of an enlightened and virtuous world that transcended their own parochial surroundings. In the lodges, the brothers cultivated a sense of superiority and exclusivity and reaffirmed their own dignity and self-worth. Freemasonry's ethos of civic responsibility accorded well with the ideology of the post-Petrine state. Russia's brothers prided themselves on their commitment to serving state and commonweal. They bore the name of Freemason as a badge of honor that distinguished them from the rest of society.

The Masons' desire to stand out from the world around them points to another major development of the eighteenth century—namely, the birth of civil society. Although it has traditionally been viewed as a country where the combination of social backwardness and political oppression prevented a civil society from developing, Russia, like other European countries at the time, was forming a civil society or "public sphere" beyond the realms of state, church, and family. In the eighteenth century, civil society referred to a number of new institutions such as clubs, literary and scientific societies, salons, theaters, and Masonic lodges. Linked with the emerging print sphere of newspapers, journals, and books, these meeting places helped unify Russia's educated classes and created among them the sense of constituting a discrete social body, a public, separate from the gray masses. Although their lodges formed a part of this civil society, the Masons adopted a paradoxical stance toward it. At once belonging to the public and ostensibly dedicated to serving society, they also saw themselves as better than society and in need of avoiding its supposed harmful influences.

The Masonic condescension toward society was in turn mirrored by society's distrust of the brothers. The Freemasons' habit of meeting in secret aroused suspicion and fear in many Russians. The Mason became a popular image of derision, mocked in the journals and on the stages of the day. Russia's brothers responded by attempting to mold nascent public opinion to their favor. They failed. Novikov's arrest attests to the great fear of Freemasonry that gripped Russian society in the final decade of the eighteenth century, when even a lone and sickly man could provoke such decisive action from the authorities. Past historians have depicted this governmental action as part of Catherine II's larger campaign against an independent society and as proof of Russian autocracy's uniquely oppressive nature.[6] Such claims are misleading, for Novikov's arrest was not ordered by Catherine, but was undertaken by the Moscow authorities acting on their own initiative. It did not signal the beginning of an organized campaign against the Masonic movement or society as a whole. Rather, Novikov's fate shows the degree to which Russia shared in the paranoid reaction to Freemasons, Jacobins, and the Illuminati that swept Europe in the wake of the French Revolution. By the early 1790s, the canard about the secret societies as sowers of the new anarchy proved too powerful for Freemasonry's defenders to counteract, and a once proud, flourishing movement collapsed, in Russia and throughout Europe, under a tidal wave of rumor and allegation.[7]

. . .

*S*hrouded in myth and obscured by legend, Freemasonry's beginnings have long been the subject of speculation and argument. As the great historian Frances A. Yates accurately put it: "The origin of Freemasonry is one of the most debated, and debatable, subjects in the whole realm of historical inquiry,"[8] and it is only recently that we have gained any objective insight into this illusive question. Built upon the foundation of the "Old Charges" or "Old Constitutions" of medieval English stonemasons (documents containing the elaborate, mythical history of the craft), the edifice of Freemasonry was first constructed in Scotland by working masons in the sixteenth and seventeenth centuries. It was the Scots and not, as traditionally believed, their English neighbors who over the span of some hundred years developed the potent mixture of lore, ritual, ceremony, and institutional structure that came to be known as Freemasonry.[9]

Although medieval masonic guilds had much in common with other craft organizations, two features set them apart. First, in order to practice their trade, stonemasons generally traveled more widely than did other craftsmen in the Middle Ages, who tended to live and work in a single place. This mobility resulted in a life of greater uncertainty for the masons, and led to the development of a special structure that supplemented the masons' conventional guild known as the lodge. Originally a simple shed, the lodge evolved into a more substantial building attached to large construction projects and served as the primary center in the masons' professional and personal lives, a place where they could gather to escape the elements, to work, rest, eat, socialize, and sleep. For masons from distant regions the lodge came to be a temporary home. Since some building projects (especially the massive cathedrals) required decades of labor, the lodges became almost permanent institutions in which the men developed their own rituals and customs to organize communal life. The masons codified ceremonies used to celebrate specific craft occasions, such as banquets and the initiation of new members, and they developed tests to distinguish the qualified mason from the impostor.

The interrogation of strangers included questions on the history of the stonemasons' craft, and this history represents the medieval masons' second distinguishing feature. The account laid down in the "Old Charges" attached supreme importance to masonry as the foundation for all other sciences. By equating masonry with geometry and then asserting geometry's preeminence as the basis for the seven classical liberal arts or sciences, a grandiose foundation myth was established that traced masonry's origins from the biblical figure of Jabal (antediluvian founder of geometry) through Hermes Trismegistus (supposed great-grandson of Noah) to the Babylonian king Nimrod, then to Abraham and Sarah, and finally to King Solomon, on whose temple some 80,000 masons labored. Some of these stonemasons later made their way to France, where they worked for the future king Charles Martel, and eventually went on to the British Isles. The epic lore of the stonemasons, their self-proclaimed status as architects and geometricians, and the pride and public recognition that came from building the great architectural structures of the age distinguished the masons from other guilds of the Middle Ages and attracted men who sought to transform the incorporation of the masons into something greater, something more mysterious and profound than had ever been known.

The key figure in this transformation was William Shaw (ca. 1550–1602), master of works to the Scottish king and general warden of all the Scottish masons, who used his powerful position to remake the traditional craft into something essentially new. Shaw created an organizational structure for the masons based on a new type of lodge; he laid the basis for the Masonic system of degrees (or grades) and the secret rituals for identification usually known as the Mason word; he also elaborated the complex symbolism associated with the stonemasons (that is, the elements of building including tools, materials, specific skills, and so on) that characterized and gave Freemasonry its distinct color. A true man of his age, Shaw was fully immersed in the intellectual currents of the late Renaissance, and it was from the milieu of Neoplatonic philosophy—and the fascination with the occult that drew learned men to subjects like magic, astrology, and alchemy—that modern Masonry was born. Onto the medieval legends of the masons, Shaw and his followers grafted various strands of the occult, from Hermeticism to the cult of Egypt, from the art of memory to the secret societies of the Rosicrucians. The use of secrecy among stonemasons (to control craft knowledge) and the predilection for symbolism were seized upon and developed according to both the Neoplatonic practice of striving for the hidden knowledge of the universe in small, secret groups and the groups' fascination with symbolism in general, particularly as expressed by hieroglyphs and emblems.

This "Renaissance contribution" is significant for two reasons.[10] First, it shows that modern, eighteenth-century Freemasonry was as much a product of the late Renaissance as it was of the Enlightenment, and that the Neoplatonic enthusiasm for secrecy, for esoteric symbolism and ritual, and for hidden truths remained a key component of Masonry throughout the eighteenth century. Thus, when in the Age of Enlightenment French, German, or Russian Masons found themselves attracted to mystical elements of the order, they were not perverting what has been perceived as Freemasonry's original, loftier, more "enlightened" impulses. Rather, they were recapturing those features of Freemasonry that went back to its earliest days.[11]

The other significant feature of the Renaissance contribution was the new category of men attracted to the lodges. Before 1600, only "operative," or working, stonemasons frequented the lodges, but in the early years of the seventeenth century these men were joined by "non-operatives," men from other social groups and occupations. These men may have been drawn to the craft by the alluring mixture of medieval masonic legend with recently added elements of the mystical and the occult that beckoned with tantalizing hints of secret knowledge and higher truths. Although some of the early non-operatives admitted to the lodges were gentlemen, dabblers in the occult sciences seeking to remake themselves into Hermetic Magi or to establish contact with the Rosicrucians, most of the new members came from more modest social backgrounds. Still, over time the lodges became increasingly popular with members of the upper classes who slowly found a place in them. Operative and non-operative masons in Scotland shared the lodges (not always gladly) throughout the seventeenth century and into the early eighteenth century, when at least thirty lodges were in operation. The arrival of non-operatives, and especially of gentlemen, was significant because modern (or "speculative" as opposed to "operative") Freemasonry is traditionally thought to have begun at the time the lodges came to be dominated by such groups. Yet, as David Stevenson points out, to suggest that Masonry only became a philosophical movement once the operative masons had been removed from its ranks is to deny such men the possibility of speculative aptitude. In fact, working Scottish stonemasons had begun to engage in esoteric rites rich in symbolic meaning before the appearance of large numbers of non-operatives among their ranks.[12]

The situation was different in England where the earliest lodges appeared to have no real links to the traditional masonic guilds and where the movement, from the beginning, consisted solely of gentlemen Masons. Lodges were established in the northern regions of the country (most likely via Scotland) by the mid-1600s, when the Oxford antiquarian and astrologer Elias Ashmole was made a Mason at Warrington in Lancashire. By the eighteenth century, England was quickly replacing Scotland as Freemasonry's new home. Free from the operative Masons' conservative resistance to innovation, the gentlemen Masons to the south changed and adapted the craft to meet their own tastes and interests. The English Freemasons added a third degree to the Scottish two, and together these three degrees became the basis of all subsequent Masonic systems. In addition, the English developed more elaborate rituals and ceremonies for use in lodge gatherings, and in 1717 they created a new organizational structure when four lodges joined to form the Grand Lodge of London, in which supreme authority over the English lodges was vested. In 1723, the Grand lodge issued the first "Constitutions of the Freemasons" to govern the affairs of all Freemasons and their lodges. By 1725, the Grand Lodge of London had more than seventy English lodges under its jurisdiction; ten years later more than one hundred lodges met in London alone. Freemasonry had become a movement of the educated and well-to-do; eventually it attracted even members of the royal family and earned the title "the royal art."[13]

Along with these changes, Freemasonry underwent a profound intellectual transformation as the world of the late Renaissance yielded to that of the early Enlightenment. Although earlier philosophical and cultural components remained evident in the Masonic predilection for hieroglyphs, esoteric knowledge, and secrecy, these elements migrated to the peripheries of the Masons' mental universe while new intellectual and social concerns came to the fore. Astrology, alchemy, and Hermeticism slowly (though not completely) gave way to modern scientific ideas, most notably Newtonianism with its sense of order and regularity. Reflecting the increasingly confident tenor of the times, the Freemasons professed an unquestioned faith in man's reason and power and committed themselves to the moral betterment of humanity. Brotherhood, toleration, social harmony, and equality became the movement's guiding principles.

These principles appealed not only to the Masons but to all of England's educated classes and formed part of the reaction to the turmoil of the previous century's Civil War since they were perceived as the best defense against future outbursts of violence. Just as pantheism and deism were meant to bar the door to dangerous religious enthusiasm, the new sociability (asserting man's natural inclination toward fellowship) and the articulation of a discourse of politeness (linking the advancement from barbarism to Enlightenment with the refinement of manners) served as effective restraints against man's more violent instincts. Along with thousands of other clubs that formed during this same time (from which Freemasonry borrowed much of its new ethos and many of its activities and goals), the lodges became crucial centers for the fostering and inculcation of these ideas. In addition, Masonry joined together men of roughly equal wealth, education, and culture and, by effacing older social boundaries, helped to instill in these men a sense of community and common identity, one that affirmed their essential equality as members of polite society and their separateness from the vulgar masses.[14]

This was the Freemasonry that the British began to export to the continent in the first decades of the eighteenth century. Merchants, along with British officers, were its chief exporters. In the foreign ports and cities these travelers visited, they set up lodges at such a rate that by 1750 Freemasonry was found in almost every country in Europe. It appears that the first lodge on the continent was established in Rotterdam in 1721 or 1722. Although the lodges were banned in the Netherlands in 1735, by the 1740s the grand lodges of England and Scotland were overseeing the creation of lodges throughout the country. By 1756, when the Grand Lodge of the Netherlands was formed marking Dutch Masonic independence, some dozen active lodges already existed.[15]

Whereas Hanoverian Britons were primarily responsible for bringing Masonry to the Low Countries, Jacobites tended to be more active in Catholic states. A group of English and Irish Jacobites are believed to have set up the first lodge in the French capital around 1725. Nowhere else on the continent did the Masonic movement flourish as it did in France, and lodges quickly sprang up not only in large urban areas but also in smaller towns along major French trade routes, in coastal settlements, and even in southern provinces. The growth of Freemasonry did not always proceed smoothly, however. During the 1730s and 1740s, the Parisian police frequently raided the lodges, arresting and interrogating the members.[16] By 1750 the lodges had spread to all parts of France, and during the 1760s more than 180 lodges were founded. On the eve of the revolution there were almost 650 lodges operating under the authority of the Grand Orient of France. Between 30,000 and 40,000 French Freemasons were active in some 1,000 lodges in the eighteenth century.[17] As in Britain, Freemasonry's attraction cut across social lines, and the lodges were filled with men from both the Third Estate, originally making up the majority of the membership, and the aristocracy, who soon patronized Freemasonry in large numbers though frequently in separate lodges.[18]

The first lodge on German soil was Hamburg's Absalom founded in 1737, most likely by English merchants. It was followed three years later by Berlin's Three Globes. Soon lodges began appearing in major trade and market cities. They spread rapidly throughout northern and southern Germany so that by 1800 almost every provincial city had a lodge. Freemasonry also made its way to the numerous petty courts via expatriate French aristocrats who introduced the movement as yet another feature of Gallic culture. It is estimated that during the course of the eighteenth century there existed between 250 and 350 lodges visited by 15,000 to 20,000 German Masons. Crucial for the lodges' popularity was Crown Prince Frederick's initiation into the order in 1738, an act that made Freemasonry socially acceptable and helped to attract titled individuals, as well as men from the middle classes, to the lodges. The movement counted among its members some of the country's leading intellectuals, such as Goethe and Lessing, and quickly became the home of the new administrative elite of the German states. Together with others from this "aristocratic-bourgeois upper class" of nobles, scholars, lawyers, and merchants, the elite found in the lodges what has been described as "an attentive sociability" and helped to promote the Enlightenment in Germany.[19]

Other parts of northern Europe received Freemasonry just as the movement was taking hold in Germany. A member of Berlin's Three Globes probably founded Denmark's first lodge in Copenhagen in 1743. The English and Scots, however, soon became the dominant force behind the growth of the lodges, and their grand lodges in turn established new lodges. In Sweden, where reports of Masonic activity date from the early 1730s, the movement, after a temporary prohibition in 1738, gained solid footing by the late 1750s. As in Germany, Freemasonry in Sweden benefited from royal patronage: in 1773, Charles, duke of Sudermania and the brother of King Gustav III, served as Grand Master of the Grand Lodge of Sweden. Two years later the king also joined the order, ensuring its continued patronage and protection.[20]

To the south, Freemasonry generally met with greater resistance due largely to the suspicion and hostility of the Catholic church, which issued three bulls against the lodges in the eighteenth century. In Spain, where English Masons had been active in Madrid and at Gibraltar since 1728, King Philip V took measures to halt the order in 1740—including condemning eight brothers "to the galleys." Such drastic punishment notwithstanding, the lodges continued to operate, at first primarily among the English and then increasingly among the Spaniards, who declared themselves independent of the English in 1767 and established their own grand lodge. Portugal had been home to English Masons since 1735, yet following the arrest and torture of the British Freemason John Coustos (the poor man was reportedly "subjected 9 times to the rack, scourged, branded, and otherwise tortured"), the Masons ceased activity until the reign of José I (1750–1777), who afforded them some degree of toleration and safety. Freemasons in the Italian states, where lodges appeared around 1735, shared a comparable fate, generally flourishing for short periods only to be persecuted and suppressed (usually at the insistence of the church).[21]

In the Swiss Confederation, a few English gentlemen established a lodge in Geneva in 1736, and three years later several of their countrymen founded the Perfect Union of Strangers in Vaud. Other lodges soon followed, but most appear to have been short-lived, leaving few traces of their existence even though as many as seventy-two lodges were in existence in Switzerland in 1787.[22] Austrian Masonry began with the initiation into the brotherhood of Francis, duke of Lorraine, the future husband of Empress Maria Theresa, in 1731 in The Hague. Vienna's Aux Trois Canons, the first lodge in Austria, was established in 1742 and was composed largely of foreign diplomats and high government officials. A second lodge (Aux Trois Coeurs) was founded in 1754 by German Masons with links to Masonic circles in Hannover. The first lodge in Prague appeared in 1726, in Hungary in the 1760s, and in Galicia around 1775, among Austrian officers.[23] Despite (or perhaps because of) the attempt to forbid the lodges in 1764, Freemasonry grew increasingly strong and popular in the final decade of Maria Theresa's reign and expanded ever more rapidly in the early years of her successor, Joseph II. In 1780, Vienna had six lodges and roughly two hundred members. This rapid growth was checked, however, by the emperor's *Freimaurerpatent* of 1785, which placed severe restrictions on the lodges and greatly weakened the movement's vitality.[24]

Freemasonry's expansion did not stop at Europe's borders. Freemasonry played too large a role in its members' lives to be left at home while they explored and colonized the world. The British colonial cities of Philadelphia, Boston, New York, and Charleston all had lodges by 1738. Eight years earlier lodges had been founded in India. Freemasonry arrived in Turkey in 1743, in Sumatra in 1765, on Java and Ceylon in 1769 and 1771 respectively, in Cape Town in 1772, and in China in 1788.[25] By the end of the century, Freemasonry was no longer an international movement: it was a global one.

As British merchants, Jacobite exiles, French aristocrats, German diplomats, and Dutch seamen disseminated Freemasonry to the ends of the earth, the movement began to undergo a seemingly unending process of mutation, reinvention, and reformation that led to a bewildering complexity of Masonic systems, degrees, titles, offices, and ceremonies. Freemasonry was nothing if not protean, as exemplified in the changes wrought by the English on Scottish Freemasonry in the late seventeenth and early eighteenth centuries. It is a long-standing though mistaken assumption (especially noticeable in Russian historiography) that early English Freemasonry represented the true, pure form of the movement, and after being exported to such peoples as the superficial French or the mystical Germans it quite regrettably became polluted and debased. Rather, it was in England where arguably the most profound and lasting reformation of Freemasonry occurred—in 1751 a group of brothers broke from the grand lodge. These brothers established their own brand of Masonry with rites and rituals that they proclaimed the authentic "Antient" Freemasonry, in contrast to the older form that came to be known (paradoxically) as "Modern."[26] Similar innovations—usually masquerading as an earlier, more authentic Masonry and almost always maintaining the constituent three grades of English Masonry—shaped the movement on the Continent as an array of competing rites arose during the century. The diverse and mutable organizational character of Freemasonry mirrored its heterogeneous intellectual, philosophical, and cultural orientations. A complex structure built on the cusp between two diverging intellectual epochs, Freemasonry lacked internal purity and consistency from the start, and this was in part the reason for its popularity. Part Medieval legend, part Renaissance Neoplatonism, part early Enlightenment optimism, Freemasonry spoke in different ways to different groups, each of which perceived a different truth embedded in the order and each of which used Freemasonry to its own particular ends.

The Masons' boast of originating in the impenetrable mist of antiquity, of possessing a timeless wisdom carefully handed down over the ages, and of practicing primordial rites and rituals belies the essential novelty of Freemasonry and points to its status as a quintessential invented tradition.[27] It is the movement's novelty and the originality at its core, artfully camouflaged in antiquated garb, that has come to attract scholarly attention over the past decades and has resulted in disagreement over how best to assess Freemasonry's historical significance.

Increasingly, historians seek to highlight the lodges' importance for the modern world by detecting in Masonic gatherings the very origins of modernity itself. For some, the lodges appear to be the matrix of political extremism, the source of antidemocratic excesses carried out by militant minorities in the name of the greater good first articulated during the Reign of Terror of the French Revolution. For others the lodges represent the birthplace of a healthy democratic spirit, of civil society, and of our modern ideas of toleration, equality, and liberty. The former interpretation owes an intellectual debt to the abbé Augustin de Barruel, an early theorist of revolutionary conspiracy, whose *Memoirs, Illustrating the History of Jacobinism* (1797) helped to establish the connection between the French Revolution and Freemasonry. The late François Furet, a leading authority on the history of the revolution, tried to revive this connection, albeit in a modified form, through his rehabilitation of the work of the conservative historian Augustin Cochin (1876–1916). Elaborating upon Cochin's thesis on the revolution, Furet repeated the assertion that the Masonic lodges, together with other "sociétés de pensée," or philosophical societies, were responsible for the creation of a new system of political relations described as "'pure' democracy"—an abstract, artificial form of democratic practice divorced from lived social experience. The philosophical society, and particularly the lodge, became the locus of a Rousseauean "general will" and marked the birth of "the first example of collective constraint" and "the tyranny of society" that made the "all-powerful machine" of Jacobinism possible.[28] Not a conspiratorial tool manipulated by revolutionary plotters à la Barruel, Freemasonry figured as "an exemplary embodiment of the new chemistry of power . . . it embodied the origins of Jacobinism."[29] Moreover, the historical import of Freemasonry was not limited to the particular excesses of the revolution. Something more terrifying, more insidious, more powerful had been brought to life. In the repressive political machine of the Masonic lodges, Furet boldly claimed to have found "the matrix of totalitarianism."[30]

Striking as such claims may be, they were neither unique nor original. As early as the 1950s, the German historian Reinhart Koselleck had sought to make the connection between eighteenth-century Freemasonry and twentieth-century totalitarianism in *Critique and Crisis: Enlightenment and the Pathogenesis of Modern Society*.[31] With its revealing title, Koselleck's work argued that the pathologies of the present era were the direct result of eighteenth-century ideologies masquerading as politically benign moral visions. Koselleck attributed a leading role in the manufacture of such utopian schemes to the Freemasons who, in the sheltered, private sphere of the lodges, developed a radical critique of absolutism that failed to recognize their own political aims. Free from the demanding exercise of political power that would have tempered their grandiose designs, convinced of the moral superiority of their beliefs, and armed with a progressive view of history, the Masons unleashed a naive and irresponsible critique of the state that sought to transcend the world of politics altogether. It was this dangerously utopian vision hatched by groups like the Freemasons that Koselleck saw as responsible for the modern totalitarianisms of both Left and Right.

These dire appraisals contrast with those of scholars who see in the lodges not the origins of radical democracy or political extremism, but the origins of civil society and laboratories for fashioning a civic, reformist, and progressive political culture. The leading proponent of this interpretation has been Margaret C. Jacob. Building upon the pioneering work of Jürgen Habermas, Jacob has sought to rescue the lodges from the Furets and Kosellecks by rejecting their sweeping generalizations of Masonic practice and ideology, instead producing detailed studies of the brothers at work in the lodges. Her exhaustive research in eighteenth-century European Freemasonry has resulted in a more balanced and nuanced picture of Masonic sociability. She reveals the ultimate hollowness in the works of Freemasonry's detractors and captures the complexities and variations of Masonic activity that were in no way divorced from lived social experience but sprang from it and sought to transform it. The desire for transformation did not imply radical extremism or a rush to transcend politics. Rather, the lodges were, in

Jacob's estimation, "microscopic civil polities . . . in effect schools for constitutional government."[32] The lodges were not radical cells in which utopian ideologies were concocted and then ruthlessly imposed on an unwitting society by a band of extremists, but public spaces in which men—and some women—from various backgrounds could gather freely to discuss pressing social, intellectual, and philosophical matters and to learn the art of constitutionalism and the politics of self-government that are the basis of modern democratic polities.

What ultimately divides these two groups of critics is their understanding of the Enlightenment. Both see the Masonic lodges as embodying the principles of the Enlightenment, and they view attempts to criticize or to defend the lodges as attacks on or defenses of the Enlightenment. One is left wondering, however, whether these competing interpretive schemes do not place too much weight on the Masons' modest shoulders. Both schools of thought betray inherent shortcomings that render their broader claims suspect. First, to equate the Enlightenment with Freemasonry requires a clear, agreed-upon definition of each term. While such a definition exists for Freemasonry, it is lacking for the Enlightenment, whose precise meaning and historical legacy are a continuing source of contention.[33] As Robert Darnton observed, the Enlightenment has been so grossly inflated by its modern critics—as well as by the great legions of *dix-huitièmistes* bent on uncovering hitherto unnoticed Enlightenment(s) from Berlin to Brazil, from Petersburg to Pennsylvania—that it has effectively been emptied of all meaning.[34] Everything and hence nothing, everywhere and hence nowhere, the Enlightenment becomes synonymous with vague notions of modernity and Western civilization, and the way in which one understands these muddled concepts shapes one's attitude toward the Enlightenment. To defend the Enlightenment is to defend the modern era and the West; to criticize the Enlightenment (for its false universalism, its naive faith in reason, its racism and sexism) is to criticize—indeed, to expose—the modern era and the West. To quote Professor Darnton, "whoever has a bone to pick or a cause to defend begins with the Enlightenment."[35] So caution must be exercised when attempting to establish a correspondence between the Enlightenment (however defined) and Freemasonry. Moreover, even if one admits Freemasonry's debt to the Enlightenment (as this and most studies do), the relationship between the two, noted by Rudolf Vierhaus, was one of similarity, not identity.[36] Freemasonry could function as a forum for the nurturing of enlightened ideas, but it could also serve as an antidote to such ideas as it did for many Russian Masons fleeing Voltairian freethinking and the subversive beliefs of the *philosophes*. To Catherine the Great, who delighted in mocking them, the Masons personified both age-old superstition and the modern credulity of fashionable society. Clearly, not all Freemasons were "living the Enlightenment."

A second problem has to do with allowing present-day debates over the Enlightenment to dictate interpretations of Freemasonry. In other words, is a criticism of Freemasonry a criticism of the Enlightenment? Or, can one criticize the Enlightenment and simultaneously look to the lodges as laudable features of the eighteenth century? It might be better to leave these contentious debates over the Enlightenment and instead to study the Masons' actions and words to discover what they reveal about the movement itself and not what conclusions can be drawn about the Enlightenment, however defined.

Finally, there is the problem of teleology, or what can be called the fallacy of origins. To seek the roots of Jacobinism and totalitarianism or of representative government and constitutionalism in the Masonic lodges of the eighteenth century is to attribute a significance to such practices that was lacking for their own participants and to do violence to the spirit in which such practices were conducted. As Margaret Jacob has rightly asked, "If the masonic lodges were the seedbeds of Jacobinism, then why did they not spawn it in Philadelphia in 1780 or in Brussels and Amsterdam in the 1790s?"[37] Masonic practice needs to be examined in its specific local context, and no one has done a better job of this than Professor Jacob. Still, it should be pointed out that, if Freemasonry was not responsible for producing Jacobinism, neither was it responsible for creating constitutionalism.[38] One needs to examine the immediate, concrete concerns of the Freemasons in order to avoid the pitfalls that spring from the desire to account for our world rather than to understand theirs. With that in mind, let us move from these larger generalities to the specifics of Russian Freemasonry in the eighteenth century, from these broad theoretical considerations to more narrowly focused investigations into the minds and actions of Russia's Masons.

1

Life in the Lodges

What an important meditation for us, dear brothers! For us Freemasons . . . people
who rose up from the world's great multitude and entered into a close union so that
in peaceful silence and with united will we might prepare our minds for wisdom
and our hearts for virtue and thus live according to our great calling.

—*Materials for Freemasons*

Understanding what attracted Russian men to the Masonic movement requires a clear idea of what precisely went on among the brothers in their
gatherings. This chapter opens the doors of the lodge and peers over the
shoulders of its members as they carried out the rites and rituals that were
at the center of Masonic life. It attempts to uncover the primary features of
Masonic practice and the concerns that animated this activity. But before we
turn our sights to the Masons' ceremonies and customs, we must first address several fundamental questions about the history and the social composition of Russia's lodges.

THE MASONIC LODGES

During the eighteenth century the Russian empire was home to more than
140 Masonic lodges located in over forty cities and towns. These lodges
formed a broad network that spanned almost the entire country and extended well into the provinces, far from any major urban centers. Russia's
initial encounters with Freemasonry date from the first half of the century.
Although no precise date can be set to mark the establishment of the initial
lodge, it is most likely that Englishmen, who were then spreading the
movement throughout Europe and beyond, brought Masonry to Russia
sometime in the 1730s or early 1740s.[1] Sources mention James Keith, a

Scotsman who fled England following the Jacobite rebellion and entered Russian service in 1728, being installed as Provincial Grand Master of all Russia in 1740 or 1741. It is possible, however, that the origins of Freemasonry in Russia go back to 1731 when the Grand Master Lord Lovell in London is reported to have appointed one Captain John Phillips as Provincial Grand Master of Russia and Germany.[2]

Freemasonry found other pathways into Russia as well. One route was via Russians who came in contact with the movement while living abroad. In 1737, S. K. Naryshkin, ambassador to England in the early 1740s, was initiated into a Masonic lodge in Paris. K. G. Razumovskii, future president of the Academy of Sciences, was a member of a Berlin lodge from 1743 to 1744 during his studies abroad as a young man. While stationed in Vienna in the 1740s, Z. G. Chernyshev, an interpreter at the Russian embassy who later became head of the Military College and governor-general of Moscow, served as secretary in the Aux Trois Canons lodge. Razumovskii and Chernyshev remained active Freemasons upon their return to Russia; thirty years later, both men were members of the Nine Muses lodge in St. Petersburg. Continental Europeans were another source for the importation of Masonry: the Swiss baron Heinrich Tschudi, an actor at the St. Petersburg court before becoming the private secretary to Empress Elizabeth's favorite, I. I. Shuvalov, helped introduce to Russia a form of French Freemasonry around the middle of the century.[3]

Despite these early encounters, it was not until after 1750 that the founding of lodges truly began to take hold. By 1770, lodges were operating in port towns such as Riga and Arkhangel'sk, and seven lodges had been established in St. Petersburg.[4] The following two decades witnessed a dramatic increase in Masonic activity: between 1770 and 1790 more than one hundred lodges were established and the total number of active lodges rose from fewer than fourteen at the end of the 1760s to approximately ninety prior to 1790. St. Petersburg and Moscow had the greatest concentration and were the key centers of Masonic activity. Initially, St. Petersburg led the way, having thirty lodges by 1780 (twenty-three dating from the 1770s), compared to Moscow's seven. But during the 1780s, the locus of Masonic activity shifted to Moscow, where almost three lodges formed for each new lodge established in the imperial capital.[5]

Even though these two cities functioned as Masonry's most important locales, more than half of all lodges were located in dozens of provincial cities and towns, particularly in ports where expatriate foreigners introduced their native habits and customs. By 1800, Riga had eight lodges, Reval (Tallinn, Estonia) had four, Arkhangel'sk had two, and Libau (Liepaia, Latvia) had one. Russia's expanding western borderlands represented another fertile site for the Masonic lodges, approximately thirty of which operated in the territories acquired through the three partitions of Poland. As might be expected, the eastern, less developed and less populous portions of the country provided a markedly less hospitable environment, and although Freemasonry's reach eventually extended as far as Irkutsk, the number of towns east of Moscow and St. Petersburg that had lodges remained low. Lodge sites east of Moscow numbered fewer than ten: Vladimir, Iaroslavl', Vologda, Nizhnii Novgorod, Penza, Simbirsk, Kazan', Perm', and Irkutsk. There were lodges in Tula and Orel to the south of Moscow; Khar'kov in Ukraine; the naval garrison on the island of Kronstadt; to the north in Petrozavodsk near the shores of Lake Onega; and Arkhangel'sk on the White Sea. While most of the provincial localities supported only a single lodge, over a dozen counted two or more. Most of these were situated along Russia's western fringe: Mitau (Jelgava, Latvia) had three lodges, for example, and Vil'na (Vilnius, Lithuania), six. Finally, there was the so-called military lodge, founded by or operating within a specific military unit and not necessarily linked permanently to any given place. During the First Turkish War (1768–1774), military officers and physicians stationed in Iassy formed the lodge Mars (about 1772), and roughly two years later officers in Sadogury in Moldavia founded the lodge Minerva. According to the memoirs of S. N. Glinka, the military lodges played a vital role in building an esprit de corps among Russian officers by instilling in them a greater sense of mutual obligation or fraternity and of undying loyalty to the Empress Catherine II.[6]

The creation of military lodges reflects the often close connection between the growth of the state and the Masonic movement. Just as Russian military officers brought the culture of Russia's capitals with them to the far reaches of the empire, so did state officials serving in the expanding provincial administration. The provincial capitals of Orel, Zhitomir, Simbirsk, and Iaroslavl', for example, all hosted Masonic lodges.[7] The lodge in Iaroslavl' was established and headed by A. P. Mel'gunov, the Iaroslav and Vologda governor-general; and Z. Ia. Karneev, the vice-governor of Orel, founded and led that town's Rising Eagle.

The majority of Russia's lodges had relatively short lives. While the exact dates of operation for a great many of them are still unknown (despite Andrei Serkov's monumental efforts), it appears that roughly one-third operated for a year or less and that about a half closed down within five years of opening. Still, over one-quarter of Russia's lodges remained in operation between six and fifteen years, and more than a dozen survived for longer than fifteen years becoming well-established social institutions in their respective cities and towns. Vil'na's Lodge of the Good Shepherd, for instance, remained active for close to forty years, and St. Petersburg's Urania (or Muse Urania) lodge flourished for more than twenty.

Lodges typically met in private homes usually belonging to a brother

Mason. In 1775, for example, the lodge Equality gathered at the residence of its brother Prince M. M. Shcherbatov, a historian and writer living in Krasnoe Selo east of Moscow. This lodge convened at other locations as well, including in early 1777 the dacha of the lodge's brother N. I. Buturlin in Moscow's Iamskaia Sloboda. The Three Banners lodge met in the 1780s at the Moscow home of its brother Friede.[8] Beginning in July 1774, St. Petersburg's Urania gathered in turn at the residence of L. Krabbe, the lodge's steward, and at the Grand Master V. I. Lukin's home. After Lukin sold his house, Urania moved its meetings in August 1774 to the Moika Canal, where the lodge rented a building for 270 rubles a year. Finally, on 8 October 1782, Urania celebrated the opening of its own residence.[9] Of all the city's lodges, Constancy was perhaps the most fortunate, for it received its own building in the early 1760s as a gift from Peter III, himself a Mason who operated his own lodge at Oranienbaum.[10] In June 1777, Baron de Corberon, the French chargé d'affaires in St. Petersburg, was initiated—"sans cérémonie"—into the seventh degree of P. I. Melissino's Masonic system, together with his friend Count Charles-Adolphe Brühl, a secretary in the Prussian legation, in Corberon's office.[11] Still, most Masons in the capital came together in private homes. Except for a single 1783 meeting in the Anichkov palace, the Grand English lodge (also known as the Provincial lodge) appears to have held its quarterly meetings in I. P. Elagin's residence on the eponymous island. The same location was used in the early 1770s by Nine Muses, which Elagin had founded and where he served as Grand Master. Elagin was also the Grand Master of the Pelican of Charity lodge, which convened for a time in the home of its member V. P. Kokushkin, a St. Petersburg merchant.[12]

The situation in Russia's provinces was similar to that in the two capitals. On 3 December 1784, for example, I. P. Turgenev founded Golden Wreath in his Simbirsk home, and in the 1780s the True Patriotism lodge gathered on the estate of Count F. F. Pototskii in Tul'chin.[13] In the summer of 1783, the Golden Key lodge of Perm' met in the apartment of its Grand Master, I. I. Panaev, Perm's future provincial procurator. This latter arrangement seems to have been less than satisfactory and presented certain obstacles to the conduct of the lodge's affairs. An entry in the minutes of the Golden Key notes that the brothers celebrated "with all the usual rites, [as much as the conditions allowed them to do so]."[14] Of all the meeting sites surely none was as unusual as the ship *Rostislav* on which the Kronstadt-based lodge Neptune met four times.[15]

The size of each lodge varied greatly. At one extreme were Rosicrucian and other highly selective lodges that sought to keep their numbers low. Certain meetings of the brothers of the Theoretical Degree, for example, were limited to nine members. At the other extreme, almost three hundred brothers of Urania and visitors from other lodges gathered in St. Petersburg

Prince M. M. Shcherbatov (1733–1790) was a historian, publicist, and administrator. Freemasonry's influence on Shcherbatov is most evident in his unfinished utopian novel *Journey to the Land of Ophir* (1784).

in the spring of 1784 to honor the passing of a fellow Mason. Other special assemblies of the Urania lodge attracted between 120 and 250 brothers. Not counting visitors who frequently attended meetings, most lodges had between a dozen and fifty members.[16] Lodge membership fluctuated and was rarely stable for any lengthy period because so many Masons were in state service and consequently were often forced to move.

THE MASONS

Russian Freemasonry has traditionally been viewed as a movement limited almost exclusively to the nobility or gentry. Recent research, however, has produced a much more complex and diverse picture of Masonry's social composition. The more than three thousand Freemasons active in Russia's lodges in the eighteenth century came from a broad range of social stations, ranks, and occupations that extended well beyond the boundaries of the privileged orders.[17] Although the gentry did dominate Masonry in sheer numbers, they never exercised a monopoly over the lodges, which remained open to all groups that formed the Russian public.

Approximately 1,100 Masons came from the ranks of the officer corps of the army and navy, and from the state civil service.[18] Allowing for the high degree of shifting back and forth between military and civil service then common, this group was almost evenly divided between the two branches of service with a slight majority serving in the military. The bulk of these Masons were born into the nobility and held a rank *(chin)* within the Table of Ranks. Nevertheless, even this fairly homogeneous group exhibited a considerable degree of diversity.

At one end of the spectrum were representatives of Russia's most powerful families—the Golitsyns, Naryshkins, Saltykovs, Trubetskois, and Vorontsovs—who dominated the highest ranks of the military, the state administration, and the imperial court, and formed the upper stratum of what one specialist has called Russia's "ruling class."[19] More than fifty members of these families were active Freemasons, including the brothers N. I. and P. I. Panin and I. G. and Z. G. Chernyshev, K. G. Razumovskii, and Prince A. B. Kurakin.[20] Approximately sixty Masons (from these and a few other prominent families) possessed a rank within the first three classes of the Table of Ranks, a position that placed them within the highest echelon of the Russian political and social elite.[21] At the other end of the spectrum were men from a much different background, such as E. F. Groot, a bookkeeper for the Riga city court, or F. L. Zande, a native German who in 1777 worked as a forest warden at the imperial country estate Izmailovo near Moscow. Totaling approximately three hundred, the number of these Masons was several times larger than that making up Russia's ruling elite. But the largest contingent of military officers and state officials lay between these two poles and consisted of men drawn from the middle and lower classes in the Table of Ranks. Of the eight hundred or so Masons with identifiable positions in the Table of Ranks, more than 750 of them served in classes four through fourteen (the lowest class); the bulk (approximately 460) came from classes four through eight—the "political core" of the country's ruling class. The entire state apparatus was thoroughly populated with Freemasons, and the Masonic lodges were filled with representatives of the tsarist state's expanding administration.[22] This interweaving of the state apparatus and Freemasonry becomes even more pronounced when one considers that scores of Freemasons from the fields of medicine, education, and the arts were also employed by the state.[23]

After the class of military officers and state officials, the next largest occupational group in the lodges comprised some three hundred merchants. They were joined by representatives of what might be called other "free professions" (that is, bankers, lawyers, manufacturers, innkeepers) who, when added to those active in military and civil service, account for almost three-quarters of all Masons with identified occupations.[24] The remainder of the Masonic community was generally evenly divided among individuals from a wide array of occupations, professions, and social groups. From the world of education and letters there were teachers, professors, students, tutors, writers, librarians, publishers, and even a geometrician; from the world of art and music there were engravers, sculptors, architects, actors, composers, musicians, and prompters. There were men who administered to the health of the body (physicians, surgeons, dentists, and pharmacists) and of the soul (Protestant ministers and Orthodox priests). Among the Masons were also found dozens of craftsmen including jewelers, dyers, and smiths, tailors, tanners, and watchmakers. A rung lower on the social scale were those involved in menial labor: gardeners, cooks, valets, and servants. Even former serfs joined the lodges. After receiving his freedom, Ivan Rodimonov was initiated into the third degree in Moscow's Three Banners lodge, which operated under Rodimonov's former master, P. A. Tatishchev. In 1786, A. N. Voronikhin, a serf belonging to A. S. Stroganov, who went on to become a prominent architect and painter, gained his freedom and joined St. Petersburg's Perfect Union.

Clearly, part of Freemasonry's attraction for these brothers of lower social standing was the opportunity to socialize with men of power and prestige. In an era in which personal relationships played a crucial role in making one's way in the world, some men must have been attracted to the order by the chance of establishing connections with individuals who might help them secure a new position or office or provide smaller favors. The precise role Freemasonry played in the exercise of patronage, how it might have worked in conjunction with (or possibly have mitigated against) the extensive clan networks to distribute power, influence, and spoils, is a complex and fascinating question that remains largely unexplored. David Ransel has drawn attention to the fact that many members of the clientele network surrounding the imperial councilor and foreign minister Nikita Panin belonged to the lodges. Important figures included I. P. Elagin, generals A. I. Bibikov, A. P. Mel'gunov, and N. V. Repnin, princes G. P. Gagarin and Aleksandr Borisovich Kurakin (as well as his brothers Aleksei and Stepan), Count A. I. Musin-Pushkin, and the writer I. F. Bogdanovich, along

Poet, dramatist, and personal secretary to Catherine II, I. P. Elagin (1725–1793) was a leader of Russian Freemasonry. Head of the country's first national Masonic union in the early 1770s, Elagin maintained a lifelong devotion to the order and was especially drawn to its mystical and occult currents.

with members of the Dolgorukii and Apraskin families. Several of these men were bound to Panin and to each other by blood and marriage as well as by state service and Freemasonry: the princes Kurakin and Gagarin were young relatives of Panin; Repnin was married to Panin's niece, Natal'ia Kurakina—whose brother, B. A. Kurakin (father of the Kurakin brothers), had also earlier enjoyed the patronage of his uncle Nikita Panin.[25]

Andrei Serkov's research makes it possible to trace the connections between the Panin party and Freemasonry even further. Just as Nikita Panin surrounded himself with Masons, so too did his principal clients. Repnin's staff, for example, included brother Masons such as Iu. A. Neledinskii-Meletskii, I. A. Alekseev, I. I. Vinter, I. N. Ivanov, E. V. Karneev, P. P. Pankrat'ev, and Count L. K. Razumovskii. While in St. Petersburg in 1776, Repnin frequented the lodge Nine Muses founded by fellow Panin client Elagin.

Head of the imperial theater administration and Catherine's personal secretary, Elagin clearly sought to combine his various roles as powerful court figure, man of letters, and Freemason to create his own patronage network connected to that of Panin. Among the members of the Nine Muses lodge, along with Repnin, were Elagin creatures such as dramatist V. I.

Lukin; imperial theater actors G. G. Volkov (brother of F. G. Volkov, the "father" of Russian theater), P. Umanov, A. Popov, P. Kozhevnikov, and N. Mikhailov; F. [P.] Bogoliubov, secretary in the imperial theater; I. S. Zakharov, former copyist under Elagin; the Italian Francesco Gradizzi, architect and stage designer at the imperial theater; composer Alessio Prati; court musicians Enrico Filipet, Giovanni Ritto, and Giuseppe Schiatto; and, finally, Elagin's son-in-law, N. I. Buturlin.[26] Early in 1773, fifteen members of Elagin's lodge established the Urania lodge under Nine Muses's supervision, and the newly chosen Grand Master was none other than Lukin, who has been referred to as "Elagin's right hand."[27] Elagin paid a visit to Urania later that year, and Lukin wisely recommended his patron for honorary membership in the lodge.[28]

While the connection between the Panin party and Freemasonry is the most easily sketched and the best known, there are two reasons not to place undue emphasis on it. First, Freemasonry was greater than patronage. It was able to transcend the intense personal loyalties produced by allegiance to the various clientele networks and numbered among its adherents men from competing clans. Indeed, the same Nine Muses to which Nikita Panin belonged in 1774 also counted as a member that very year his "long-time enemy" Z. G. Chernyshev.[29] Second, Panin was not the only powerful figure to attract the company of brother Masons, as a glimpse at Chernyshev shows. While governor-general of Moscow between 1782 and 1784, Chernyshev patronized the local Freemasons and employed several of them on his staff, including S. I. Gamaleia, I. P. Turgenev, and Prince N. N. Shakhovskoi. On a more personal level, Johann Kölchin, a Freemason born in Riga, served as the doctor in Chernyshev's home, and I. M. Strazhev, a former serf who had earlier belonged to Mogilev's Hercules in the Cradle, together with S. I. Gamaleia, worked for a while as the manager of part of Chernyshev's estates.

A striking diversity of national and ethnic origin mirrored that of social rank and standing, and almost all of Europe was represented in Russia's lodges: England, Scotland, France, Spain, Denmark, Sweden, Holland, Austria, Germany, Poland, Italy, Hungary, Greece, and Armenia. The single largest group was made up of ethnic Russians, followed by Germans (mostly from Russia's Baltic provinces), and then Polish subjects of the empire. Most of the Russians and Poles who filled the lodges were nobles, while the German and English Masons generally came from bourgeois backgrounds. The languages in which meetings were conducted—Russian, English, German, French, and even Italian—reflected Freemasonry's cosmopolitan nature. Some lodges operated in one official language, but those with more mixed membership often used several. The Urania lodge in St. Petersburg, for example, began using only Russian in 1773, before adding German two years later; it dropped Russian in favor of English during the

1780s, when the number of Russian members dwindled. Riga's Small World, on the other hand, in its initial meetings in the winter of 1790, operated only in German (as did most Baltic lodges) and then added meetings in Russian beginning in the autumn.[30]

National and ethnic diversity brought with it differences in religious affiliation as well, and although this diversity was limited primarily to Christianity—that is, Russian Orthodoxy, Lutheranism, and Roman Catholicism—it did include a handful of Jews. On 16 August 1788, for instance, St. Petersburg's Urania lodge admitted two Jewish merchants: Isaak Levin from Potsdam and Moses Oppenheim from Königsberg. They rose quickly in the lodge's hierarchy, and by the end of the month the former had been initiated into the fifth Masonic degree and the latter into the third. Apparently, the large donations that both made upon their initiation facilitated this noticeably swift advancement.[31]

The heterogeneity of the Masonic movement was not always evident in the membership of a specific lodge, and not surprisingly certain groups chose to segregate themselves in a particular lodge. Nine Muses had many members drawn from the spheres of literature and the arts, while guards officers predominated in the lodge Bellona. Moscow's Union of Foreigners consisted almost exclusively of expatriate Frenchmen. The urban lodges in which foreigners predominated were often composed primarily of merchants—like the Holy Catherine lodge in Arkhangel'sk—and to a lesser extent of military officers and state officials. Local nobles tended to preponderate in the Baltic lodges, whereas more than half of the members of one Scottish Rite lodge in Moscow were Russian princes. There is also evidence to suggest that a large group of workmen imported from Scotland in the 1780s by the architect Charles Cameron to help with the construction of the empress's summer residence at Tsarskoe Selo constituted an "Imperial Scottish Lodge of St. Petersburg." If this lodge did in fact meet, then it would have represented a rarity within eighteenth-century Freemasonry: namely, a lodge consisting of operative masons (actual stonemasons) and not the non-operatives found in other Masonic lodges.[32]

Nevertheless, what is most striking about the composition of the lodges is their degree of diversity of rank and station, a fact that did not escape the brothers themselves, like the orator of Moscow's Three Banners lodge who commented that "the order contained in its composition a great number of members from all estates."[33] Georg Reinbeck, a German traveler to Russia in the early nineteenth century, provided a similar assessment of the lodges' social makeup. According to Reinbeck, the movement had been wildly popular with men from a wide range of social backgrounds, a state of affairs that (with an unmistakable air of national condescension) he saw as a regrettable corruption of Freemasonry at the hands of the Russians: "The nation . . . embraced Freemasonry with enthusiasm, but the object of the society was perverted. Everybody was admitted without scrutiny for the sake of the fees. Still this had one good effect, that of bringing the different ranks of life nearer to each other. At last the rage of masonry increased to such a degree, that the empress often saw her court deserted, even by the gentlemen in waiting, and when she asked where they had been, the constant answer was, 'At the lodge.'"[34] Representative of this diversity was Moscow's Astreia lodge (active in 1783), made up of military officers, state officials, surveyors, merchants, a physician, a translator, a student, a bookkeeper, and a professor. Of those occupied in military service, the lodge's members included brigadiers (V class), a second lieutenant (XIII class), and two sergeants; those in civil service ranged from the level of state counselor (VII class) all the way down to Senate junior clerk.[35] Along with the former serf Strazhev, Mogilev's Hercules in the Cradle lodge included among its members A. I. Verevkin, the director of economy (VI class) for the Mogilev province; V. I. Polianskii, a local official (VII class) who had studied and traveled in Europe where he had met Voltaire; and J. G. Schwarz, founder of the lodge and tutor in the home of A. M. Rakhmanov (a fellow Mason and chairman of Mogilev's criminal court).[36]

Of course, one important group remained generally absent from the ranks of the Freemasons: women. The lodge was an almost exclusively male preserve that barred its doors to what the Masons perceived to be the distracting presence of the opposite sex. This prohibition notwithstanding, some European women did participate in the Masonic movement in what came to be known as *maçonnerie des dames* or *maçonnerie des femmes*, now usually referred to as adoption lodges. These lodges first appeared in France as early as the 1740s, and it was there that they achieved their fullest development and greatest acceptance. In 1774, the Grand Orient of France officially recognized *maçonnerie des femmes*, and by the eve of the revolution adoption lodges were found in almost every large French city and in many towns. Although *maçonnerie des femmes* was especially popular among the aristocracy and members of the *haute bourgeoisie*, it also attracted women of more humble station, such as actresses from The Hague's Comédie Française who belonged to the local Loge de Juste. According to Janet M. Burke, a leading authority on French *maçonnerie des femmes*, the adoption lodges made a profound impact by offering elite women a place in which to imbibe Enlightenment notions of fraternity and virtue, and, to a lesser extent, liberty and equality—ideas, she notes, that shaped both their private and their public behavior. Despite its title, *maçonnerie des femmes* actually included both sexes, and women Masons were never allowed to meet without the supervision of their male counterparts.

Nevertheless, during the second half of the century women Masons gained greater control over their Masonic affairs, as the rise of a brand of female Masonry that included a higher degree called *Amazonnerie Anglaise*, or

Ordre des Amazonnes, wonderfully shows. Women Masons of this degree not only began to organize their own rituals and to run their meetings with greater independence, they even began to question male authority outside the lodges. As Janet Burke and Margaret Jacob have observed, the English Amazons' catechism went so far as to "call on women to recognize the injustice of men, to throw off the masculine yoke, to dominate in marriage, and to claim equal wealth with men." While the *Ordre des Amazonnes* must be seen as an extreme wing of Freemasonry, when viewed together with *maçonnerie des femmes* as a whole it raises serious doubts about recent claims of Freemasonry's purported misogyny and its representativeness of a supposedly masculine Enlightenment intent on expelling women from the public sphere. The research of Burke and Jacob reveals a decidedly different picture and demonstrates that Freemasonry was not anathema to women's interests but could actually provide a forum for their cultivation. In the adoption lodges, hundreds of European women encountered the ideas of the Enlightenment, and the lodges even became places "where large numbers of women first expressed what we may legitimately describe as an early feminism."[37]

Not surprisingly, women's Masonry proved to be controversial, and there is little evidence that it ever became widespread outside France. According to Andrei Serkov, three adoption lodges operated in the Russian empire: Three Crowned Hearts, established in Mitau in the spring of 1779; Absolute Loyalty, active in Vil'na from 1780; and Diffused Darkness, founded in Zhitomir in 1786. Unfortunately, little is known about these lodges. Vil'na's Absolute Loyalty, headed by Countess Przeździecki (née Radziwill) and composed of local aristocratic women, appears to have worked in conjunction with that town's all-male Complete Harmony lodge, several of whose members also belonged to Absolute Loyalty. The origins of Three Crowned Hearts have been linked to the Courland visit of the great eighteenth-century adventurer Cagliostro. The inventor of his own "Egyptian" brand of Freemasonry that was open to both sexes and that featured his fetching wife, Lorenza, as Grand Mistress, Cagliostro apparently founded Three Crowned Hearts as part of a larger strategy to ingratiate himself with the leaders of local society. From his new brother and sister Masons, Cagliostro hoped to acquire the necessary letters of introduction for a pending trip to St. Petersburg where, it has been suggested, he planned to set up an adoption lodge that would include among its members none other than the Empress Catherine II.[38] Cagliostro's initial contact in the capital was the Baltic German baron Karl-Heinrich Heyking, a colonel in Russian service and an active Freemason. Transferred to Warsaw not long after Cagliostro's arrival, Heyking immersed himself in that city's Masonic affairs where he helped to establish an adoption lodge, complete with special insignia, ceremonies, and degrees, and named after Hypatia (ca. 370–415), the Neoplatonist philosopher, mathematician, and astronomer murdered by Christians in Alexandria. According to Heyking, this lodge comprised "ladies of the highest society and of outstanding civility" and proved to be a great success.[39]

There exists one more tantalizing piece of evidence concerning the possible existence of female Masonry in Russia. In the introduction to his popular *Proofs of a Conspiracy against all the Religions and Governments of Europe, Carried on in the Secret Meetings of Free Masons, Illuminati, and Reading Societies,* first published in Edinburgh in 1797, John Robison relates his encounters with Freemasonry during his stay in St. Petersburg in the early 1770s. According to Robison,

> my masonic rank admitted me to a very elegant entertainment in the female *Loge de la Fidelité,* where every ceremonial was composed in the highest degree of elegance, and every thing conducted with the most delicate respect for our fair sisters, and the old song of brotherly love was chanted in the most refined strain of sentiment. I do not suppose that the Parisian Free Masonry of forty-five degrees could give me more entertainment. I had profited so much by it, that I had the honour of being appointed the Brother-orator.[40]

We know that Robison frequented St. Petersburg's English lodge Perfect Union, and it is conceivable that this Loge de la Fidelité, if it in fact existed, operated in conjunction with it. Sadly, there is nothing to corroborate Robison's intriguing statement.

MASONIC PRACTICES

Most Masonic lodges met two or three times a month.[41] The Masons convoked different types of meetings, usually called "lodges" in Masonic parlance, depending on the work at hand. Most common of all meetings were initiation lodges intended for the induction of new members or the advancement of a brother Mason into the next degree. These meetings took on the name of the degree into which the brother—or brothers—was being initiated: new members joined the order during Apprentice lodges; Apprentices rose to the second Masonic degree during Fellow Craft lodges, and so on. The brothers discussed and voted on the lodge's general affairs in conference lodges. While day-to-day issues were usually taken up in Apprentice conference lodges (that is, open to all brothers regardless of degree), matters deemed more pressing and serious were addressed in conference lodges open only to the lodge's officials and those at the level of the Master degree.[42] During instruction lodges the brothers received lessons on the history of the order, learned the meanings attached to various Masonic rites and symbols, or listened to edifying speeches delivered by one of the brothers.[43]

A lodge of sorrow, during which one of the lodge brothers gave a short eulogy, marked the death of a brother Mason. According to Masonic law, three separate lodges of sorrow for each brother were to be held—two in the same month of his death, the third in the following month.[44]

Many lodges, especially initiation lodges, concluded with a banquet: the brothers would gather around the table to share a large meal, to sing Masonic songs, to listen perhaps to a brief didactic speech, and, most important, "to fire their guns"—that is, to drink toasts, though ideally in moderation.[45] Any gathering, therefore, could consist of one or more "lodges." On 26 April 1774, for example, Urania opened its proceedings with a "Master lodge" during which two Fellow Craft brothers were initiated into the Master degree; after being officially closed, an "Apprentice conference lodge" was then opened at which time the lodge's Apprentices and Fellow Craft degree members were admitted into the meeting.[46] In addition to these typical gatherings, the Masons also came together to celebrate special occasions. Most notable of these celebrations was John the Baptist Day on June 24, when all Masons honored the order's patron with a full day of special rites and ceremonies, beginning with a church service, then speeches, poetry readings, a short outing to a local park or garden, and concluding with an especially extravagant banquet.[47]

Lodge meetings usually convened at either 5:00 or 6:00 in the evening, depending on the time of year—the later hour generally preferred during the summer months. Each meeting began with a brief opening ritual, followed by the reading of the Masons' general statutes and the minutes from the last meeting. Next the brothers turned to the work at hand—be it initiating a new brother, listening to a morally instructive speech or exegesis of Masonic hieroglyphs, or addressing the broad range of mundane administrative concerns that constantly came up, such as raising enough money to purchase the necessary lodge adornments or, as occurred once at the Urania lodge, deciding what to do with a fifty-ruble cello given to the lodge as a gift.[48] Upon completion of their work, the brothers made donations to the lodge's benevolence fund, the receipts from which were then counted and the sum duly recorded in the minutes. The meeting ended with a short prayer and closing ceremony, at which point the brothers either departed or sat down to a banquet.

A group of elected officials administered each lodge, and although there was some variation among Russia's lodges, most of them had a comparable organizational structure based on seven constituent officers. At the top of the lodge hierarchy was the Master, or Grand Master, who held supreme authority and ran each lodge meeting. Next came the Senior and Junior Wardens, who aided the Master in bringing the lodge brothers to order and in conducting the many rites and ceremonies. Together these three formed the nucleus of every lodge: their presence constituted the "formation of the lodge" and in their absence no lodge could officially convene.[49] The Secretary was charged with the lodge's correspondence, with reading aloud the Masonic statutes and regulations before meetings, with making occasional speeches, and, most important, with keeping minutes in which the actions and words of every meeting were to be recorded faithfully. The primary duties of the fifth official, the Orator, included preparing all candidates for the order and for the higher degrees by questioning them at length about their reasons for seeking admission and by testing their understanding and knowledge of Masonry. Finally, each lodge had a Treasurer, who oversaw all the lodge's monetary affairs, and a Master of Ceremonies, whose many duties included determining the exact ceremonial forms appropriate for banquets and other celebrations.[50]

Along with these officials, most lodges had a few other recognized positions established to guarantee their smooth operation. A guard stationed at the entrance admitted only lodge brothers holding an entrance pass and approved guests. Servant brothers acted as factotums carrying out a range of menial tasks, such as serving food and drink during Masonic banquets. Among the ten servant brothers active in Kronstadt's Neptune lodge were three musicians who performed during banquets, Fellow Craft initiations, and special celebrations. In addition, most lodges had alms collectors who gathered regular donations for the benevolence fund.[51]

· · ·

*E*very lodge possessed a set of laws and statutes without which it could not be officially recognized by other lodges. One of the first steps in establishing a new lodge was the acquisition of laws and statutes from an existing lodge having the authority to supply them. These documents, usually consisting of Masonic regulations, bylaws, and detailed instructions necessary for various rites and ceremonies, set the framework within which the Masons operated. About 1780, the Rising Sun lodge in Kazan' received from Moscow's Provincial lodge a collection of official documents in four light-blue books written in a neat hand and emblazoned with the Provincial lodge's bright red seal. Perhaps the most complete extant set of eighteenth-century Russian Masonic documents, these books include the fifty-two General Laws regulating all aspects of lodge life, especially the words, actions, and responsibilities of the brothers. Largely overlooked by past scholars, these General Laws provide an excellent point of entry into an examination of the Masonic world.[52]

The laws are striking in their scope and specificity, ranging from such broad matters as the necessity of loving all brother Freemasons throughout the world to much more particular concerns like the proper place for carrying one's Masonic badge. As the lodges were self-financing, the laws naturally devote a good deal of attention to ensuring a steady flow of income.

Charter issued by Grand Provincial Master I. P. Elagin establishing the St. Petersburg Nine Muses lodge in 1772. Statutes, laws, charters, and constitutions were necessary components for Freemasonry's proper and orderly functioning. Only Grand lodges had the authority to issue charters to members enabling them to establish officially their own lodge and to confer degrees.

Thus, except for the few deemed worthy but admitted for free because they lacked financial resources, most new members had to pay a fifteen-ruble initiation fee. They were also required to make gifts of no less than fifty kopecks to each of the servant brothers, and well-to-do initiates were encouraged to make an additional monetary contribution. Similar rules governed advancement to the higher degrees: admission to the Fellow Craft degree cost fifteen rubles with double that amount charged for becoming a Master degree Mason.[53] In addition to contributing to the lodge's benevolence fund, every brother also had to pay monthly dues set by each lodge to cover operating expenses. And brothers who attended any of the special celebrations were required to pay their share of the expenses in advance.[54]

Especially pronounced in the General Laws is the intent to regulate the Masons' comportment during lodge meetings. Concern was expressed about ensuring quiet and decorum. Law §8 prohibited "any unnecessary walking about, slamming of doors, and loud talking," and law §19 stipulated that during an "initiation, vote, or any other work" no brother could leave the lodge or even "change his seat" without permission. During their work (*work* being the word the Masons used to describe their activities), no whispering or talking was allowed, or "any impropriety, either in word or deed, of whatever nature." Should a brother arrive drunk at a lodge meeting or become inebriated during a banquet, his brothers were bound to conduct him carefully home, making sure that "no outsiders noticed this vice in him." Brothers were forbidden to take part in "crude jokes, rudeness, laughter and ridicule of a brother, or of an uninitiated person," and as might be expected, "quarrelling and . . . fighting in the lodge" were strictly prohibited. To ensure order during discussions, anyone who wished to speak had first to request permission from the Wardens before taking the floor. At that point, none of the brothers could interrupt unless permitted by the officers.

Besides discouraging the development of tensions and disagreements among the members, the laws sought to foster a sense of community and mutual respect. Regardless of the social distance that separated the Masons in the outside world, once inside the lodge they became members of a common fraternity and thus "brother" was the sole appellation that could be used there. Harsh penalties were set for those who said something about another brother that might engender "anger, hatred, enmity, an argument, or a falling-out."[55]

Just as strict rules were set to control the brothers' behavior in the lodge, so too were limits established on what they could talk about while there. No one, "regardless of who he may be," was allowed to discuss matters of religion or "state affairs and disagreements." Taking the Lord's name in vain or the utterance of any "devilish name" were also punishable offenses. Along with these specific prohibitions was a broader injunction against all "caustic remarks."[56]

But the numerous rules and regulations meant to dictate the conduct of the Masons were not intended to operate solely within the confines of the lodge, and the General Laws evince an equally strong commitment to directing their behavior in the outside world:

> Every brother is obliged to lead a pious and chaste life, pure in soul and body, and to try throughout his entire life and in all his deeds to display decency and fairness: he must be charitable to his neighbor, and especially to his brothers; a noble composure is that special goal towards which each Mason strives; he endeavors to hold sway over his sensual and spiritual, or mental, passions. . . . In short, in all his actions, thoughts, and feelings he must make an effort to be a light before others, and through his own example to disgrace and put to shame every vice.[57]

In other words, the same high standards expected of the brothers inside the lodge were to extend to all spheres of social life so that the Freemasons might everywhere serve as examples of "gentleness, humility, and chastity."[58]

These dozens of rules represented more than mere admonitions, and the General Laws dictated an elaborate system of fines, penalties, and surveillance to guarantee compliance. First of all, infractions resulted in monetary fines whose amount depended on the severity of the transgression. Simple failings—forgetting one's entrance pass or the lodge badge—brought a five-kopeck fine; more grievous errors, such as continuing to speak after the Master struck his gavel, unexcused absence from a scheduled lodge meeting, or swearing, cost ten kopecks, fifty kopecks, and one ruble respectively. Those who through their "lack of abstemiousness" became inebriated at a lodge banquet had to pay a five-ruble fine for the first infraction and ten rubles for the second. Equally harsh treatment was meted out to those who were unable to hold their tongue: forgetting the law against discussing religion or state affairs in the lodge resulted in a five-ruble fine. The highest fine (twenty-five rubles) was levied against those who divulged to a "false brother or to a completely unenlightened individual" what had taken place in a lodge meeting. The "brother collectors" gathered all fines and deposited them in the lodge's benevolence fund. If a brother refused to pay, the fine doubled. Any disagreements between the collectors and the fined brother were to be settled first by the two Wardens; if they failed to reach an agreement satisfactory to both parties, the matter went to the lodge Master who could decide it on his own or place the issue before the judgment of all the members.[59]

Other infractions resulted in either temporary or, in extreme cases, permanent expulsion. Brothers who abused their status by taking advantage of those seeking admission to the movement or of those in the lodge with a lower degree (that is, accepting money to aid initiation or advancement to a higher degree) ran the risk of being suspended for at least three months; those who failed to follow the decisions of the lodge Master or two Wardens were also suspended for three months, and permanent expulsion from the order faced those who exhibited "rudeness" toward the Master. In general, however, the laws left a great deal of discretion to the lodge officers. Thus, reports of serious transgressions, such as bringing a non-Mason to a meeting or continued episodes of drunkenness, could be passed on by the lodge officers to superiors in charge of their particular Masonic system who would then rule on the appropriate punishment.[60]

Enforcing compliance with Masonry's General Laws during lodge meetings presented no significant challenge. Yet since the laws were intended to guide the Masons' words and actions beyond the lodge as well, a mechanism of mutual surveillance had to be constructed:

> Every brother having witnessed in the behavior of another brother, either inside or outside the lodge, any impropriety that could possibly bring shame . . .

upon him and the brotherhood is obliged to inform immediately his lodge's Grand Master and Wardens for his own rescue and to demonstrate his loyalty to the brotherhood; the alms collectors and all officers as officials are especially bound to keep watch continually over the brothers' behavior inside as well as outside the lodge.[61]

The laws required brothers who had observed such failings or had any complaint to lodge against another to inform quietly either one of the Wardens or the Master and to await their decision "patiently in utter silence."[62]

It is impossible to tell to what extent the brothers of the Rising Sun lodge followed the rules for proper comportment. The General Laws and other statutes were simply normative texts, and the minutes of this particular lodge—the key document that would have recorded any disciplinary problems—have not survived. Such documents from other Russian lodges do offer examples of the ways in which laws and regulations affected the lives of Masons. The minutes from St. Petersburg's Urania, for example, provide numerous instances of brothers being punished for behavior deemed unbecoming a Mason. After a recommendation by Urania's Master V. I. Lukin during a meeting on 21 December 1773, the entire lodge agreed that brother Stepan not only be prohibited from participating in the next banquet, but that he must "kneel at the door" during their meal as punishment for "his drunkenness and disgraceful conduct" at a recent gathering of the Grand Provincial lodge. Almost twenty years later, in November 1790, the lodge suspended for a few months four brothers guilty of creating disturbances during meetings. Not all of the Urania's brothers were as fortunate. In 1788, the servant brother Schmidt was banished from the lodge after numerous complaints about his deportment. A decade earlier, in February 1775, the lodge's treasurer, a certain Melle, was permanently banished from the order for expropriating lodge funds; his "foolish deed" was reported to the Grand Provincial lodge as well as to all allied lodges, and lodge brothers who happened to see him were instructed not to talk with him about Freemasonry. Even visiting brothers from other lodges could be prohibited from ever returning to Urania: one Nikolai Pomorskoi who, having "both during and after the banquet committed so many indecencies that even to mention them is unpleasant," was henceforth barred from all lodge gatherings. Urania's officials considered this decision, which had been reached during a Master lodge, so important they made special note that it be announced before a meeting of the entire lodge.[63]

Other lodges were similarly devoted to monitoring the behavior of their members. A letter sent to Provincial Grand Master I. P. Elagin in December 1788, complained that two brothers of the newly established lodge Hope of Innocence in Reval had behaved badly and had delivered indecent speeches dangerous for the entire brotherhood. In 1790, Riga's Small World admonished at least two of its members for their alcoholic excesses and negligence in fulfilling their duties, and three years later nine members of the lodge Modesty were expelled: eight for outstanding debts and one for "inappropriate and indecent speech." And finally, a membership list of the Astreia lodge from 1783 includes the gloss "expelled until improvement in conduct" next to the name of E. B. Syreishikov, a professor and Master degree Mason.[64] On at least one occasion the leadership of a lodge found itself the subject of criticism from rank-and-file members. In the summer of 1787, when a letter sent to the officers of the lodge Concord—complaining about the lack of decorum during banquets and the failure to read the Masonic laws and statutes on a regular basis—did not elicit the desired changes, four lodge members continued to press the lodge authorities for action. The officers responded by accusing the brothers of behaving in an impudent manner and had them promptly expelled.[65]

Perhaps the most disturbing incident of inappropriate behavior, both for its severity and for the status of the Mason reported to have carried it out, concerned the violent outburst of J. G. Schwarz, one of the leading figures in Moscow's Masonic circles and a professor at the local university. During a holiday outing to the woods at Kuntsevo, just west of the city, Schwarz is said to have lost his temper and in a fit of rage drawn his sword on a brother Freemason. The poor man's life was saved at the last moment by several brothers who managed to subdue the enraged Schwarz. The fact that Schwarz appears not to have suffered in any way for his grievous offense suggests that Masonic law did not apply equally to all.[66]

WORKING THE ROUGH STONE

As the discussion of Masonic practice suggests, Freemasons perceived their actions as possessing great significance—both for themselves and for the rest of society. They saw themselves as engaged in nothing less than the construction of new men. Their official goals stressed "improving [our] very selves" by remaking each Mason into "a useful fellow man" and an "example of piety and virtue."[67] Virtue, defined as "the capacity or inclination of the soul to act in accordance with natural law and the principles of faith, which oblige man to fulfill his duties in relation to God as well as to himself and his neighbor,"[68] could be attained through what Russians of the period called *nravouchenie*, or "moral admonition." This translation is somewhat misleading, however, since it fails to capture much of the area of meaning marked out by the term. As W. Gareth Jones has noted, in the second half of the eighteenth century the idea of *nrav* (pl. *nravy*) contained in *nravouchenie* corresponded less to the English concept of "morals" than to the broader French notion of "mœurs," usually rendered as "morals and manners."[69] The *Dictionary of the Russian Academy* (1793) defined *nrav* as: "1) A natural or acquired propensity, an inclination to something good or depraved . . . 2) A habit, custom introduced by usage in any people; a manner of behavior adopted by any people."[70] *Nravy*, therefore, implied a manner of bearing—either good or bad—that was at once both "inborn" and yet capable of being acquired or learned. It is this latter meaning that is expressed by the notion of *nravouchenie*, defined in the same source as a "field of knowledge containing within it instructions, rules leading to a virtuous life, to the curbing of one's passions, and to the execution of man's responsibility and duties."[71]

The Freemasons professed to know the "path to virtue" through their possession of the fundamental rules of moral admonition.[72] They portrayed the lodge as a "school teaching the rules by which man attains knowledge of nature, himself, and God" and claimed that by adopting their "life-saving rules" man would be led inescapably to virtue and happiness.[73] Indeed, as the following excerpt from the Master degree initiation ceremony shows, Masons were constantly reminded of the primary significance of these rules:

> Having sensed your imperfection and weaknesses you joined our society with the intention of improving yourself and making yourself more complete. From the moment of your joining, we have tried to inculcate within you the rules of true virtue, penetrating to your heart by all the paths known to man.[74]

The specific "rules of true virtue" were laid out in several standard texts, most notably the Statutes or Regulations of the Freemasons and the Exegesis of the Statutes or Regulations of the Freemasons, an expanded and more thorough statement of the principles set forth in the former document.[75] The nine specific rules contained in these parallel texts are presented as a series of obligations and duties. First among them is the Mason's obligation to God and religion as the foundation of all life. The Freemason is exhorted to "fulfill all the moral obligations" of God's law and to exhibit in all his actions "enlightened piety without hypocrisy and falsity, or fanaticism."[76] Next come the Mason's duties to his earthly sovereign and to his homeland. The Statutes command tersely that every Mason "Honor the state's ruler. Love the fatherland. Fulfill with complete exactness all the duties of the good citizen."[77] In much greater detail, the Exegesis sets forth the Freemason's obligation to be "the bravest soldier, the fairest judge, the most good-natured gentleman, the truest servant, the most affectionate father, the most constant husband, [and] the most devoted son."[78]

Following his obligations to God and country, the Mason's subsequent duties were to humanity in general and to his neighbor specifically. Since Masonic doctrine contended that everyone originated from a common divine source, it followed that all humanity was worthy of the Mason's love and respect; the Mason was accordingly instructed to take a forceful part in advancing the well-being of his fellow man. Basing his dealings on an "[a]ctive, sensible, and universal charity," the Mason was always to be "affectionate and tender," and his actions free from any traces of "miserliness and vanity" so as to light the "fire of virtue" in every heart.[79] Despite the importance assigned to the responsibilities to one's fellow man, the Mason's duties to his brothers and to the order received special attention. The bond of brotherhood was to supplant the ties of narrow Christian confession, nationality, and social standing and to require not only that each Mason help every other brother in danger or in need, but even that he be willing to sacrifice his own life in carrying out this duty. Having made his decision to join the order freely, each member was bound to carry out Masonry's "moral obligations" and to submit fully to its authority: "Within the order your will is subject to the laws and the will of your superiors."[80]

Most important of all the rules, the very cornerstone upon which the others were built, was the personal moral perfection that came through knowledge of the self: "Go into your heart often, probe its innermost forces. Self-knowledge is the foundation of all Masonic principles. Your soul is a rough stone that must be planed. Sacrifice to God your reformed and subdued passions."[81] Just as the rough or unworked stone found in nature was marked by irregular surfaces and sharp craggy edges, so too were men's souls seen as scarred, uneven, and disfigured. The unbridled drives and passions holding sway in men's souls were seen as the cause of this disfigurement. In his natural state, man was perceived as a lost and helpless creature, entrapped by and defenseless against the vices that dictated his thoughts and actions as they deformed his soul and corrupted his morals and manners.

The view of human nature presented to new members of the order, therefore, characterized man as "having been born with passions and having grown accustomed to their delightfully wicked dominion." During initiation rites, new members were warned that "our passions and weaknesses, darkening the inner light, lead us into blindness" and were admonished to "grow out of the gloom of our passions."[82] As long as these passions remained untamed, they would hold the Mason in a state of darkness—blind to his own predicament and cut off from the liberating light of virtue. Thus, when during initiations into the order the Grand Master asked what constituted the Freemason's chief obligation, the Senior Warden responded: "To escape vice and make progress in virtue." Specifically, this meant replacing the vices of "pride, cupidity, intemperance, suspicion and hatred" with the virtues of "modesty, good behavior, caution, [and] charity."[83]

After recognizing the fact that he was enslaved by his passions, the Mason had next to subdue them, and Masonic discourse and practice repeatedly reminded him of this fact. One set of Masonic documents included among a list of the Freemasons' primary duties "[l]ove for the death of our passions," describing this demise as "the sole means to free oneself from the captivity of vices," and urging the brothers to "shackle those feelings that arouse the excitement of the passions."[84] When asked by the Grand Master during the initiation ceremony why he had come to the lodge, the Senior Warden responded, "To master my passions."[85] In fact, the appellation "Freemason" implied one who through his diligent work on the rough stone had become *free* from the grip of passions and vices.[86]

In practical terms mastering the passions meant becoming one's own watchman and keeping an honest and unblinking eye on the internal workings of the self. In a speech from the 1780s, a Russian Freemason described this operation as the requirement to "keep watch constantly over all the impulses of our mind and heart, to scrutinize their every disposition and every action. As soon as their ends become carnal pleasure or intellectual pride, then in no sense are they true and do not belong to our vow."[87]

To help stimulate and direct such self-examination, one Masonic text lists twelve general rules to be hung on the lodge wall. They include:

> **2.** Examine every evening the preceding day.
>
> **4.** Turn your attention to the history of your life. What was good and bad in my parents, in my first teachers, in my young friends, and in the schools in which I grew up? Do I really possess now those good qualities whose foundation these circumstances laid in me, or in what way have I become worse? . . . What are the most noble and what are the worst deeds in my life that I can recall?
>
> **6.** Explore your fantasy—imagine yourself in all sorts of situations and ask yourself how you would endure them. How would I act in this instance? What people don't I like, yet still consider to be good, and what part in this displeasure do my own vices play? What am I like when angry, when in love, when happy, when confronted with unhappiness, when insulted?[88]

I. P. Turgenev's draft of a personal confession most likely meant to be read at a lodge meeting shows that such instructions to inspect the inner workings of the self were taken seriously:

> As regards my body, my primary sin is intemperance and my favorite passion is voluptuousness or, rather, gluttony, for in matters of food I do not possess a refined palate. It is from this disposition to gluttony that my disposition to debauchery springs, against which I must struggle every day. I am unable to attain that level of sensation where my imagination would be free from that foul vice. This disposition is a product of my brutish soul. The gluttony, which burdens my stomach . . . aggravates the vice of my build [that is, his corpulence] and my indolence.[89]

Turgenev was no less self-critical in assessing the condition of his soul:

With regard to my spiritual passions, it seems to me that chief among them is pride, which manifests itself in the pleasure I take in listening to others' praise, especially from those persons I hold in high esteem. I know not what I would refuse to undertake or to carry out so as to merit such praise. Faint-heartedness is visible in my soul's pride. I give this name to the way I indulge my wife and family . . . I admit to this vice and as a sign of my candor I would not be ashamed to describe here in detail the indulgences I have shown her and which I, to my own shame, know so well . . . So then, gluttony, indolence, lack of self-restraint, the love of praise, indulging my wife, and faint-heartedness, which, because my pride commands that I hide it, produces hypocrisy . . . and pretense.[90]

Masonry's elaborate symbolic language, with its intricate systems of hieroglyphs and allegories, provided the medium through which these ideas and attitudes about the self were inculcated. Everywhere the Mason looked in the lodge, every text he read, every speech he heard, and every ritual in which he participated impressed upon him the importance of working the rough stone within. As part of the initiation ceremony, for example, every new brother learned the meanings behind the various symbols depicted on the Apprentice degree carpet.[91] The sun, moon, and stars were to remind the Apprentice that he "must keep vigil over his actions day and night," and the compass (representing "reason") that reason always be used to measure these actions. The trowel was a sign of the need to smooth the cracks and fissures in his heart "produced . . . by arrogance, anger, despair, weakness, and vices." In addition, each new brother received a small trowel to be worn along with the rest of his Masonic adornments as a reminder of this obligation to work the rough stone. The general aim—to continue with the Masonic metaphors—was to reshape the rough stone so that its original state became unrecognizable: no longer covered with unhewn and jagged surfaces, it was to be "scoured, planed . . . and smoothed out."[92]

The Mason's mores reflected his success, or failure, in this civilizing process: "May your manners be pure and clean, your soul righteous, honest, and humble."[93] Whereas an unworked soul would find physical expression through unrefined behavior, characterized by an intemperate and undisciplined bearing, an even, regular, and level soul would be reflected by a manner that was controlled, measured, and moderate. The planed soul then was one that strictly observed "all the rules of honesty, truth, and decency."[94] The quality and style of one's deportment communicated the purity of one's mœurs.

This fact helps to bring out some of the larger meaning and significance of the exacting rules placed on the Masons' behavior as exhibited in the General Laws. Outbursts of any kind and instances of behavioral excess were signs of a broader moral failing—symptoms of still unbridled

A Masonic carpet. Neophytes were instructed through the use of carpets depicting various Masonic symbols appropriate for that degree. This Apprentice carpet includes the tools necessary for the Freemason in his work upon himself such as the level, compass, trowel, and square. These are to be used to turn the rough stone or ashlar (shown on the left) into the perfect, polished stone (on the right). The seven steps between the two pillars of Jachin and Boaz (Wisdom and Strength) ascending to the western gate of Solomon's Temple mark the path that the Apprentice must follow.

passions, an undisciplined self, and consequently, of impure and corrupt morals. The Mason was required to be the very embodiment of pure morals, and he was reminded of this at every turn. When, during his initiation into the Master degree, the Freemason placed his legs together to form part of a rectangle, the officers instructed him to take this as a sign of the "well-behaved life" expected of every Master Mason and of the "[r]espectable precision and caution demanded of him in all of his conduct."[95]

The Mason, then, represented someone who possessed and demonstrated through his deportment a moral state of *dobronravie* (good behavior), understood here as a gentle and modest disposition. However, the idea of good behavior encompassed more than this limited notion of restraint and modesty; it also embraced the idea of an inclination to do good, to fulfill one's duties and obligations. In October 1782, S. I. Gamaleia stressed this broader meaning in his speech "On Good Behavior" *(O dobronravii)*, ad-

A Freemason Formed out of the Materials of His Lodge (1754). This illustration uses various Masonic tools and symbols to represent the character of the Freemason. The twin pillars of Jachin and Boaz form the legs. Above the Mason's apron, the body comprises two globes— terrestrial and celestial—and the Holy Bible. A level acts as the neck and shoulders to which squares have been attached for arms. Each hand has been replaced with an iron cramp called a lewis, intended to signify strength; a plumb line dangles from the left hand, and a compass rests atop the right. The head is a radiant sun from which hangs the jewel of the Past Master. On the mosaic floor sit a tracing-board illustrating Euclidean geometry and a stone, chisel, and maul. The overall impression is one of symmetry, proportion, regularity, and order—the qualities demanded of every Freemason. This illustration may well have belonged to I. P. Elagin.

dressed to the brothers of the Deucalion lodge (aptly named after Prometheus's son, who created a new race of men, the Leleges, following the retributive flood Zeus used to destroy humanity). He noted that good behavior was much more than simply being "affectionate, tolerant, gentle, charitable, [and] obliging."

> [G]ood behavior is nothing but liking what is good, loving what is good; but liking good so completely that one no longer likes evil . . . Such a person always likes good, he is always in the habit of doing good, i.e., he has become good, orderly, useful, complete.[96]

All of these elements—the need for greater restraint and modesty in word and deed, for unquestionable devotion to God, sovereign, homeland,

and one's fellow man, and for absolute loyalty to the order itself—found their highest expression in one defining moment: the Masonic oath sworn by all initiates. Before God and the entire lodge, every new member pledged his allegiance to the greater authority of church, state, country, and brotherhood and promised to try with all of his energy to be "prudent and wise in all my acts, cautious in my deeds, moderate in my words, upright in my posts, just in my undertakings, honest in my judgments, philanthropic, noble, good-hearted and full of love for all humanity in my dealings with others."[97] The newly initiated brother vowed therefore to make himself anew, to transform himself into something better and more noble.

The lodge, as a school for the inculcation of virtue, was the setting for this transformation. It offered members the specific knowledge—as set out in its rules and regulations and as practiced in its rites and ceremonies— necessary to remake themselves into new men of virtue and piety. Through obedient and dedicated submission to the Masonic laws and statutes, the Freemason hoped to escape the prison of his blinding passions and to learn to live in harmony with natural law and the principles of a broadly defined religious faith. This transformation had ramifications that went well beyond the sphere of the lodge and the brothers' personal lives. It meant that they would be better able to fulfill their religious, political, and societal duties and, through exemplary acts, to set a higher standard for all. As G. N. Neledinskii remarked in a 1791 speech on such duties to a gathering in Orel: "[H]e most particularly must serve in society and look after the temporal and eternal well-being of his neighbor . . . purifying our hearts and enlightening our minds, we have to reform the entire world with our lives and our deeds, presenting in them the example of piety and virtue."[98] In short, by working toward their own moral perfection, the Masons saw themselves as advancing the welfare of the entire commonwealth.

. . .

*I*t is important to note that the Masons' preoccupation with virtue, moral instruction, and mores in no way marked them as unusual. When Prince Shcherbatov wrote that "the greatest strength and the greatest welfare of the state is founded on good morals," he expressed a belief common not merely to his Masonic brothers, but to educated society as a whole.[99] The concerns that animated the Masonic project were prevalent throughout society and resonated well beyond the walls of the lodge. A useful place to begin an investigation into the broader intellectual background within which the Masonic movement flourished is the *Instruction* of 1767, written by Catherine II for the Legislative Commission convened to codify Russia's laws. As one scholar has observed, politics in Catherinian Russia were to a large extent considered "a matter of civic morality," an idea that the *Instruc-*

tion expresses quite clearly.[100] The text assigned central import to "Virtue," "Good Manners," and "Morals" in the maintenance of social stability and effective governance of the polity. Of the nine specific "Objects for the Preservation of Good Order" (also referred to as "Police"), upholding "Purity of Manners" among the populace ranked second in importance, just after safeguarding the conduct of religious services and just ahead of protecting the general health of the nation.[101] Especially threatening to the commonweal were acts thought to undermine the overall condition of manners and morals; consequently, offenses "contrary to Good Manners" formed one of the four separate classes of crimes outlined in the *Instruction*.[102] Reflecting the enlightened sources upon which it was based, from Montesquieu's *Spirit of the Laws* to Beccaria's *Treatise on Crimes and Punishments*, the *Instruction* insisted that prevention should outweigh punishment. This in turn implied two direct consequences. First, government "should take more Care to instill Good Manners into the Minds of the Citizens, by proper Regulations, than to dispirit them by the Terror of corporal and capital Punishments," and second, virtue should be rewarded as a deterrent to crime. While held to be the more difficult to achieve, the former option was deemed the more efficacious. The ultimate program for regulating the polity, then, had to be based upon inculcating manners and morals, on creating and implementing the right "System of Education" that could "mend the Morals of the People."[103]

In light of this, it is not surprising that education in Catherinian Russia focused particularly on the moral development of the country's young. One sees this in the educational policies of I. I. Betskoi—the empress's chief aide in the sphere of education, who stressed the importance of implanting proper manners and of shaping youths' malleable souls—and in the chief primer of the period, *On the Duties of Man and Citizen*.[104] Divided into four principal parts devoted to such matters as "the care of the body" and "social duties," the text began with a section on "the education of the soul," which detailed the vices to be avoided, the attitudes and behavior that reflected proper morals and true virtue (for example, "rectitude," "peace of mind," and "uprightness"), and the importance these held for helping each individual carry out his duties to God, neighbor, and self.[105] The education of the soul—the pursuit of virtue through the inculcation of good manners and morals—was not a matter appropriate only for children, and throughout the century, moral admonition was taught at Moscow University as one of the three primary branches of philosophy, along with metaphysics and logic. The university expected students not only to study proper morals and manners but also to possess them, and written confirmation of "good behavior" was added as a requirement for admission in the 1770s. Other educational institutions also emphasized moral instruction. In Kazan', for instance, Julius von Canitz, the local school director, delivered a seven-year series of public lectures on topics intended to aid "the formation and improvement of morals."[106]

The general preoccupation with the improvement of morals and manners is further evident in the age's fondness for all manner of "rules for behavior." Catherine's legislation exemplifies this phenomenon. The "proper Regulations" for instilling "Good Manners" mentioned in the *Instruction* made their appearance in the Police Ordinance of 1782, which included the "Rules of Good Behavior" and the "Rules of Social Obligations" demanded of all the empire's inhabitants. Eleven terse regulations, based on the golden rule and on a broad, catholic Christian morality that mirrored the ideals found in the precepts of the Masons, instructed Russians on how they were to treat their neighbors and fellow citizens as well as what was expected of them as husbands, wives, parents, and children. The requirements for state officials exhibited moral qualities similar to those the Masons sought to inculcate in the lodges: "1) Common sense. 2) Goodwill in the exercise of commissions. 3) Love of one's fellow man. 4) Loyalty in the service of her Imperial Majesty. 5) Dedication to the common good. 6) Zeal for duty. 7) Honesty and impartiality."[107]

Even less official, more informal settings demanded explicit rules on proper comportment. At Catherine's intimate evening parties in the Hermitage, the invited guests were punished if they infringed on any of the gathering's established guidelines: for example, "Leave all hats, walking-sticks, and ranks outside the door; don't have a sullen expression; don't get overly excited and don't argue."[108] Guests at Prince A. B. Kurakin's estate Nadezhdino (Hope) near Saratov were presented with printed rules of etiquette that they were obliged to follow during their stay.[109] Perhaps the best proof of the pervasiveness of such rules and regulations was their appearance as an object of satire in the latter part of the century. Writing in the *Observer* in 1792, I. A. Krylov lampooned the public's appetite for concrete rules on correct deportment—and the common obsession with mimicking the latest behavioral styles it implied—by providing seven simple rules on how any young man ought to act if, whatever his true intelligence and character, he sought to appear witty and learned in polite society.[110]

A leading source for guidelines to appropriate behavior, indeed to all questions pertaining to manners and morals, was the so-called edifying literature *(nravouchitel'naia literatura)* that enjoyed great popularity among Russia's readers. Beginning with the publication of *The Honest Mirror of Youth* in the reign of Peter the Great, the country's reading public showed a pronounced appetite for didactic literature relating to manners and morals. Over the course of the eighteenth century, more and more men and women wanted to know what exactly constituted decorous comportment. How could they acquire the proper habits and attributes? And what was the relationship between polite manners and a virtuous soul?[111] Readers sought answers to such questions in journals and books, many of them translations, with titles like *The Honorable Man's Pocket Book, or Useful Maxims for Every Place and Every Occasion; Friendly Counsel to a Young Man Beginning to Live in the World;* and *Duties of the Honorable Man.* In these and scores of comparable texts, Russians read about the importance of modesty, of being pleasing yet honest and never deceitful, of keeping one's passions and emotions in check, and of associating as much as possible with groups and individuals respected for their integrity and virtue while avoiding persons of known depravity.[112] Although focusing primarily on the amelioration of the individual, edifying literature did not fail to highlight the larger social importance of self-improvement: by fashioning himself into an honest man *(chestnyi chelovek)*—someone who was dependable, truthful, and industrious—the reader learned that he was simultaneously improving society.[113]

This literature depicted models of virtue on which its readers were to pattern themselves and furnished precise and detailed rules, maxims, and instructions on exactly how to achieve this noble end. Russia's Masons played a prominent role in the publication and dissemination of such literature through journals like *Morning Light* (1777–1780), *Moscow Monthly Edition* (1781), and *Evening Glow* (1782), publications traditionally described as "Masonic" because their publishers were well-known Freemasons, but better seen as part of the genre of edifying literature. As the publishers of *Morning Light* pointed out, only articles believed to be "the most effective for the inculcation and establishment of good morals and manners" were selected for their journal, and to this end they fed their readers a healthy diet of morally uplifting material—again, largely translations—with titles like "Seneca's Discourses on the Honest Life, in which is Taught how to Live as a Good Man" and "Discourses on a few Passages selected from Tacitus."[114] In these and other pieces, Russians received instruction in proper morals and manners and in how to define the true character of virtue. Yet, significant as these journals were for the dissemination of such knowledge, by no means were they unique in their intellectual orientation: *Monthly Compositions for Profit and Amusement* (1755–1764), *Idle Time for Good Use* (1759–1760), and *Companion of Lovers of the Russian Word* (1783–1784), to give a few examples, were all devoted to providing readers with the most beneficial and entertaining moral instruction.[115] Even the writer and publisher I. P. Pnin, representative of that part of Russia's intellectual and political spectrum that was distinct from and often opposed to Freemasonry, published translations and penned articles in the *St. Petersburg Journal* on moral admonition and on the necessity of freeing oneself from the prison of the passions in order to become a true and worthy citizen.[116]

Virtue and manners and morals, then, occupied a position near the center of political, cultural, and intellectual life in eighteenth-century Russia. They were deemed the primary foundation upon which to construct both the body and the body politic. They formed part of the mental frame through which the educated classes viewed the social order, and they constituted a few of the core categories used to construct the contemporary world. Their significance was evident to everyone—from empress to provincial school teachers (and their pupils), from powerful court aristocrats to the growing ranks of curious readers scattered throughout the country.

The importance attached to this complex of ideas can be accounted for by laying bare its relation to several interconnected historical developments and intellectual and cultural expressions: the rise of the absolutist state, the tenets of moral philosophy, and the discourse of civility. As Gerhard Oestreich has shown, the powerful, centralized European monarchies of the seventeenth and eighteenth centuries, with their brilliant courts, massive standing armies, and large bureaucracies—perhaps best exemplified by Louis XIV's France (1643–1715)—were brought into being by and in turn reinforced a new type of man, one characterized by a strong sense of self-control, moderation, obedience, and duty. Oestreich located the source of this new man, whom Charles Taylor has aptly called the "'punctual' self,"[117] in Neostoicism, a set of ideas originating in the sixteenth century that helped make the absolutist state possible by providing the justification for the dramatic expansion of state power and the disciplining of society. According to Oestreich, Neostoicism's "aim was to increase the power and efficiency of the state by an acceptance of the central role of force and of the army. At the same time, Neostoicism also demanded self-discipline and the extension of the duties of the ruler and the moral education of the army, the officials, and indeed the whole people to a life of work, frugality, dutifulness and obedience."[118] The Neostoic man exhibited "steadfastness, patience, firmness"; he placed greater worth on "inner values" over "material goods"; he was devoted to serving the community—the *patria*—and believed that to be a good man was to be a good citizen; he possessed a "pious yet active faith, and genuine reverence for God."[119] Drawing largely upon the writings of Seneca and Tacitus, Neostoicism had become the new faith of the educated classes in France and the Netherlands by 1600, and in time Stoic ethics established themselves as the recognized moral creed in most of Europe.[120] Their influence lasted throughout the seventeenth and eighteenth centuries, shaping the thought of such philosophers as Blaise Pascal, Baruch Spinoza, Hugo Grotius, and Samuel von Pufendorf.

Freemasonry's debt to Neostoicism is undeniable. The focus on self-control and self-inspection, on moderation and discipline, on loyalty to the ruler and the state, and on service to the *patria* and one's fellow man, all reflect Neostoicism and its ethos. Indeed, *Morning Light's* translations of Seneca and Tacitus provide direct textual evidence of this influence, and the testimony of the Masons themselves shows a clear understanding of Freemasonry's power to instill such values in its members. When, in 1777,

V. Ia. Aleevtsev, the head clerk of the Belorussian governor's office, was called before the governor and archbishop of Mogilev and ordered to justify his Masonic activities, he did so by demonstrating the order's ability to turn drunken, lazy, and unruly ne'er-do-wells into sober, diligent, and orderly state officials. Aleevtsev contrasted the typical "rogues and troublemakers" who filled the governor's office with those men who had joined the society: the latter were the ones quietly seated in a row hard at work at their desks, their clothes clean and tidy, their hair neatly done in braids.[121] This incident, which apparently made strong Masonic supporters of the governor and archbishop (who even asked if he might send Aleevtsev a few rowdy monks for reforming), captures the logic of social disciplining that characterized the "well-ordered police state" of the early modern period and that Neostoicism and Freemasonry helped to make possible.[122]

The roots of Neostoicism's appeal and, concomitantly, the longing for a strong, effective state lay in the religious and political conflicts of sixteenth-century Europe, conflicts that produced a desire for peace, order, and stability.[123] The continuing upheavals of the following century not only assured the lasting attraction of Neostoicism as an antidote to a disordered and chaotic world, they also—along with the destabilizing growth of scientific knowledge—served to guarantee a prominent place in the intellectual life of the next two centuries to moral philosophy in general, of which Neostoicism formed an influential part. Compelled to reexamine traditional notions of human nature and of man's place in society, moral philosophers like Nicholas Malebranche, Anthony Ashley Cooper (the third earl of Shaftesbury), Francis Hutcheson, and Christian Wolff worked to establish a "system of morality" that would serve as the basis for moral conduct. Simultaneously a description of human nature and a set of normative standards, moral philosophy sought to show the ways to attain virtue and happiness, to define good character and proper comportment, and to delineate one's responsibilities to God, sovereign, and one's fellow man.[124]

Answers to such questions were sought with special urgency in a seventeenth-century England profoundly shaken by the political, social, and religious upheavals of the Civil War. Crucial to preventing a repeat of the excesses of that era was the necessity of restraining the passions and enthusiasms, especially religious ones, that were perceived as responsible for the chaos. Part of the answer lay in curbing men's wilder impulses and making them into reasonable, sociable, and restrained beings, a process that one scholar has called the "Reformation of Male Manners."[125] At the end of the seventeenth century, reformers launched a campaign against all forms of crude, cruel, and exaggerated behavior that saw the establishment of societies devoted to the elimination of bad morals and manners. By 1700, London had roughly twenty such organizations, the best known of which was the Society for the Reformation of Manners, founded in 1692. The rough and hardened mores of the English proved a more tenacious foe than their opponents anticipated, however, and the campaign, after suffering numerous defeats, had to be renewed again and again throughout the eighteenth century.[126] Even though they failed to achieve the complete amelioration of manners, and barbarous behavior did not become a thing of the past, the reformers, with the help of moral philosophers like Shaftesbury, at least succeeded in proving the importance of manners for the smooth operation of society. A consensus was achieved that saw in manners and morals the keys to a peaceful, harmonious, and prosperous social order, one in which the dangers of zealotry, crude dogmatism, and blinding passions were safely restrained and man's baser drives harnessed in the interest of more socially beneficial tasks.

Freemasonry was another expression of this search for a new system of morality that characterized moral philosophy and the reformation of manners. In the lodges, Freemasons worked to fashion their own moral code based on a deeper awareness of the self that would guide them to a virtuous existence and the fulfillment of their many familial and social obligations. While much of this code was indebted to Neostoicism, it could not be reduced to that alone, for the preoccupation with making oneself reasonable, sociable, and polite bespoke a set of concerns and ideas common to moral philosophy in a wider and more general sense, concerns that extended well beyond the relatively circumscribed ideas of Neostoicism.

The pronounced desire to perfect one's manners, to smooth one's rough edges, reflected another logic as well. By "working the rough stone" the Masons sought to "civilize" themselves, to become polite, refined, and agreeable. Here the discourse of Freemasonry became part of the more general discourse of civility or politeness with its attention to polishing man's crude nature, to shaping him into something smooth, level, and even. Jean Starobinski's discussion of the semantics of "civilization" in the late seventeenth and eighteenth centuries throws into bold relief the striking similarities between these two discourses:

> Owing to the association of the "literal" image of brightness and smoothness with the idea of perfection, the manual act of polishing *(expolitio, exornatio)* establishes a figurative equivalence between "to polish" and "to civilize." To civilize, whether men or things, is thus to flatten all rough edges, to exclude all possible sources of friction, to make sure that all contacts are easy and smooth. The file and polishing stone are the instruments that make possible the figurative transformation of crudeness and rusticity into civility, urbanity, and culture.[127]

Starobinski's characterization—the images and metaphors he uses—would have been readily understood by the Freemasons. A 1780 article in

Morning Light entitled "Edifying Principles" used the same language to describe the process of self-improvement: "People correct themselves, they reform their morals and manners through everyday social intercourse just as ice-floes are rounded off and become smooth from striking against each other."[128] V. N. Zinov'ev made a comparable observation in his travel diary while visiting Manchester in 1786:

> The kindness and civility of the people of this land are excellent, and I cannot speak too highly of them and attribute my little deserved happiness to them . . . I think that a person trying to learn his deficiencies could find no better way to correct them than through travel, just as a stone, which by rubbing against a great number of others becomes smoother and, so to speak, polishes itself, sometimes evens the roughness of those that strike against it.[129]

Zinov'ev's use of the image of the stone was probably more than mere coincidence since he was himself a Mason, having joined a lodge in Berlin in 1784 and been active in Lyon's Masonic circles several years later.[130] Like his fellow Masons, Zinov'ev saw in the order a way to transform himself into a model of civility and politeness through the acquisition of a new mode of being characterized by more refined manners. The necessity of such a transformation was never in doubt given the general consensus about the coarseness, even barbarism, of Russian mores. This was painfully apparent to Russians who had spent time abroad and were able to compare their countrymen's customs with those of the rest of Europe: "Our mores are so deformed," Zinov'ev wrote in St. Petersburg in 1790, "that nowhere else are such things done as they are here."[131]

The barbarism of Russian mores was a common theme among European observers. The memoirs of the Frenchman Charles François Philibert Masson, who spent ten years in Russia during the reigns of Catherine II and Paul I, are indicative of the general disregard for Russia's inhabitants, whom he described as exhibiting "depraved manners" and a "barbarism . . . displayed in vulgarity of manners."[132] Masson's opinions were typical of the age and reflect a belief that peoples and nations could be defined, categorized, and ranked based on the condition of their manners. There were, as Louis, chevalier de Jaucourt, wrote in the *Encyclopédie*, "polished nations" and "barbarous nations," and there was little doubt to which Russia, with its "Asiatic manners," belonged.[133] This did not mean, however, that Russia was destined to remain stuck in Asia, for along with this classificatory scheme arose the idea that peoples could in fact improve their manners, that they could become polite and refined, and so lift themselves out of barbarism and into civilization. In fact, Russia showed signs of already being on its way to this higher level, as the French ambassador to the court of Catherine the Great, Count Louis Philippe de Ségur, observed: "The aspect of Petersburg strikes the spirit with a double astonishment; there are united the age of barbarism and that of civilization, the tenth and the eighteenth centuries, the manners of Asia and those of Europe, coarse Scythians and polished Europeans."[134]

Freemasonry offered Russians the opportunity to shed their "Asian" manners for those of the "European," to smooth their barbaric coarseness into a civilized polish. True, Russia's Freemasons had to be careful not to take this civilizing process too far—for to appear too refined, too polite and polished, implied sacrificing morals for manners and confusing being with seeming. It was the possibility of succumbing to this danger that infused such power into the image of the overly Frenchified fop or dandy (the *petimetr* or *shchegol'*) who forsook his morals in the rush to adopt European manners.[135] The key lay in finding the proper balance between the two, for manners did matter a great deal and could not simply be ignored. Although members of the Russian public feared being labeled as laughable fops who, through their mindless and mechanical mimicking of European fashions, were perceived to be nothing more than soulless automatons, neither did they wish to be considered crude boors lacking the proper social graces. No one, not even those who berated their compatriots for selling their invaluable Russian souls for the most recent imports from Paris, wished to appear as coarse rubes or as social interlopers ignorant of the ways of society. Even Prince Shcherbatov, who in *On the Corruption of Morals in Russia* (1787) issued a blistering indictment of the deleterious consequences of European influence on Russian mores, acknowledged that Peter the Great had been right to "soften the crude ancient manners and morals" of Muscovy by introducing Western mores.[136] In fact, Russians of Shcherbatov's era had so completely internalized the discourse of civility—with its constituent contrasts of coarseness and refinement, rusticity and urbanity, civilization and barbarism—that it had become a permanent filter through which they viewed the world, yet whose presence escaped them. Moreover, the logic of civility contained within itself the idea that morals and inner virtue possessed external forms, that they made themselves known by way of a pleasing manner.[137] The Masons approached this problem from both ends, refining their manners as they reformed their morals. If the Freemason exhibited a deportment more decorous and agreeable than usually observed in society, then this was nothing more than the true, visible expression of his elevated virtue and morals—the ultimate essence of his being that he had so carefully cultivated amid his brothers within the sanctuary of the lodge.

CHAPTER

2

The Russian Public

Or, Civil Society in the Eighteenth Century

In June of 1769 the following announcement appeared in the St. Petersburg weekly the *Drone:*

> To appease his contemporary *All Sorts of Things,* the publisher of the *Drone* did not want to print this letter, but in all fairness he could not refuse Mr. Truthlover to do so, especially as *All Sorts of Things* has submitted him to the verdict of the public; thus, the sensible and dispassionate readers may settle this proceeding as they see fit, in due form or not. Herewith follows the justification of Mr. Truthlover.[1]

This brief excerpt formed part of that year's lively debate between the *Drone* and *All Sorts of Things,* a well-known episode in the intellectual and cultural life of Catherinian Russia that has traditionally been interpreted as a daring confrontation between the young journalist and publisher Nikolai Novikov and the Empress Catherine II. But as W. Gareth Jones has recently revealed, this view of the debate is more than simply anachronistic; rather, it is the product of a very conscious historical manipulation by nineteenth-century critics who, through the creation of the myth of "a personal clash between the autocrat and the tribunal of the people," sought to politicize Russia's past for their own contemporary social and political goals. Its participants were not locked in a polemical battle fraught with momentous historic implications, but were merely engaging in a discussion about the appropriate use of satire, a discussion that, true to the literary norms of the day, was punctuated with dramatic displays of moral outrage, intellectual disdain, and feigned umbrage.[2]

Jones's skillful uncovering of the true nature of the debates of 1769 has direct bearing on the present investigation since the supposed titanic struggle between Novikov and Catherine is enmeshed in the mythology that surrounds (and distorts) the history of Russian Freemasonry. According to the established view, their initial confrontation was an unmistakable portent of the Novikov tragedy and led inexorably to his arrest and imprisonment in the Schlüsselburg fortress in 1792. Historians have interpreted the suppression of Novikov as part of the empress's larger campaign to destroy the Masonic movement and even society itself. While such a view is appealing in its simplification of history's complexity to an easily comprehended story of the personal clash between good and evil, truth and power, this scenario is problematic on several levels. First, along with the fact that these literary debates lacked any hidden political undertones, there is no evidence that *All Sorts of Things* was Catherine's mouthpiece and not that of its editors; second, this view exaggerates the empress's involvement in the Novikov affair and diminishes the key role played by Prozorovskii, the local governor-general;[3] third, and most important, although Novikov's position was crucial in both Freemasonry and society, his arrest cannot be equated with the destruction of either of these institutions.

The last point brings up a larger question that has escaped past students of the lodges—namely, what exactly was the relationship between Freemasonry and Russian society? It is common to speak of Russia in the eighteenth century as having a "polite" or "educated" society. But what is meant by "society"—be it polite, educated, or civil—in eighteenth-century Russia? The answer to this question is crucial for understanding the relationship between Freemasonry and Russian society, and for assessing Freemasonry's place in it. The lodges did not exist in a barren social landscape but were part of a larger network of new institutions. In order to gain a greater awareness of the environment within which Russia's lodges operated, one must adopt a broader vantage point from which to sketch the topography of Russian society in the eighteenth century.

Toward a Definition of the Russian Public

One might start by returning to the passage that began this chapter. In the rush to create the myth of a battle between Catherine and Novikov, nineteenth-century critics overlooked the explicit reference in these debates to a third party, that is, "the public." As the excerpt shows, the publisher of the *Drone* recognized that the disagreements between the two weeklies— expressed in this instance through the persona of Mr. Truthlover—would be best resolved by their "sensible and dispassionate readers" whose collective opinion constituted the "verdict of the public."

In this appeal to the public, one sees the general recognition of some idea of society—*publika* (the public) being the century's preferred label for what may also be called polite or educated society. While the existence of a public has long been recognized by observers of Russia, the paucity of research on this rather elusive body may be due to the broadly held notion that even though Russia possessed a public, that public remained captive to the all-powerful tsarist state, impotent and underdeveloped, a mere shadow of the "real" publics found in more "advanced" western European countries like England and France.[4] Whatever the objective differences that distinguished imperial Russia from the rest of Europe, one must not succumb to this view of the Russian public as insignificant but, rather, ask if this perception is truly accurate.[5] As a growing body of work is beginning to show, such assumptions about Russian society do not hold up under scrutiny. Indeed, over the course of the eighteenth and nineteenth centuries there developed in Russia an increasingly vibrant public and public opinion whose place and role in the country's history are only now being investigated.[6]

The Russian public, as Nicholas Riasanovsky observed, was the creation of the reforms begun under Peter I and continued by his successors.[7] More specifically, the public was the product of a complex set of interrelated processes, most notably rapid Westernization, the growth of the absolutist state, and the development of economic and commercial activity dating from the latter decades of the seventeenth century and continuing through the eighteenth. Modern notions of "the public" and "public opinion" first took hold in other parts of Europe at approximately the same time. Largely on the basis of the pioneering work of Jürgen Habermas, scholars in a variety of disciplines have engaged over the past decade or so in an investigation of this "public sphere" and its import for rethinking our understanding of Old Regime societies. Focusing on such topics as the emergence of a network of new associative forms (for example, salons, coffeehouses, learned societies, and so on), the fashioning of a novel sociability within them, as well as the development of a market for periodical literature and other forms of print culture, these numerous studies provide a powerful set of heuristic tools for examining Russia's own public sphere or civil society.[8]

Central to the development of the public throughout Europe was the rise of the absolutist state that, through its permanent administrative apparatus and standing army, signaled the establishment of a "sphere of public authority."[9] In Russia, as mentioned above, this process of state formation became apparent during the reign of Peter I, when the Russian state began to take on the salient characteristics of European absolutism. Evidence of this transformation can be seen in the appearance of such terms as "the public" (*publika*), "public" (*publichnyi*), and "publication" (*publikatsiia*) in official Russian documents of the first quarter of the century.[10] In one of the first recorded uses, the General Regulation of 1720 distinguished between "public State affairs" and "private affairs." The novelty of these terms is conveyed by their inclusion in a glossary of foreign words at the end of the text in which *publichnyi* (public) is defined as *vsenarodnyi* (national; nationwide) and *privatnyi* (private) as *osobyi* (particular or individual).[11] In another early example, a source from 1730 used the expression "in the public" (*na publike*) as synonymous with "in the state" *(na gosudarstve).*[12] The "public" referred, therefore, to a new, depersonalized sphere anchored in the bureaucratic offices and institutions of the expanding and increasingly activist absolutist state. The appearance of this "well-ordered police state," whose history Marc Raeff has masterfully recounted, marked a sea change in the very notion of the state and its role in governance. Founded upon a new set of assumptions about the nature and function of political authority, the absolutist state was no longer satisfied merely to limit and control undesirable behavior and actions as had its Muscovite predecessor; rather it set out to effect positive change, to unleash the creative but untapped energies of society and thereby further the general good.[13] The state authorities as constituted by these emerging institutions and practices were deemed public because their function was the improvement of the collective welfare. Examples of such official public bodies include the Legislative Commission of 1767 and the Board of Public Welfare established in 1775. Although the idea of the public as an expression of the official, governmental sphere of the central authorities remained intact, by the second half of the century a significant transformation in its meaning had manifested itself. The "public's" field of reference expanded to include the *addressee* of state authority as well: that is, the social body that the state sought to regulate, stimulate, and control through its complex of administrative practices.[14]

Ivan Nordstet's two-volume *Russian Dictionary with German and French Translations* (1780, 1782) offers evidence of this transformation. Here *publika,* for which *obshchestvo* (society) is listed as a synonym, is translated by the French "le public" and the German "das ganze Volk" and "das Publicum."[15] Somewhat surprisingly, given its adoption into the language in the early part of the century, *publika* is missing from the *Dictionary of the Russian Academy* published in the 1780s and 1790s. In this source, under the term "society" *(obshchestvo),* which Nordstet had defined as synonymous with "the public," two primary definitions are given: "1) A people living together under the same laws, under known statutes, [or] regulations . . . 'Man is born for society. Man is obligated to be useful to society' . . . 2) An estate of people; an assembly of many persons bound together by an identical purpose or a common aim. 'A society of learned men. A merchant society.'"[16] Society *(obshchestvo),* therefore, could refer to the entire social body or populace of a particular community or to a smaller, distinguishable group within that larger collective. Nordstet's dictionary gives similar definitions: the German and French equivalents for society are listed as "die Gemeinde, Versammlung . . . Gesellschaft . . . la communauté, société, le commun" for the former meaning and "eine Gesellschaft von gelehrten Leuten, une société de gens de lettres" for the latter.[17]

Although Nordstet's dictionary defined the public as synonymous with society, the area demarcated by the former term does not appear to have been as broad as that expressed by the latter. Rather, the social referent of the public corresponded more closely to society's secondary meaning as a distinct subset within the larger social body. In this sense the Russian notion of the public shared important features with the German *Publikum*, which went through similar semantic shifts during the century. Whereas an earlier *publicum* could refer either to the state sphere or to all of society, by the middle of the century *das Publikum* (having acquired around this time its Germanized form) came to signify a completely new "social community" separate and distinct from the state (as in *der Staat* or *das Gemeinwesen*) and the common folk (as in *das Volk* or *die Gemeinde*). After 1750, *das Publikum* became a "fashionable catchword of the upper classes": the public formed "a social stratum of the educated classes, 'society,'" whose "taste and manners" distinguished it from the masses.[18] The new meaning developed out of the former primary usage of *das Publikum*—as an audience, a body of spectators, listeners, or readers—that had arisen earlier with the appearance of a host of new cultural forms such as concerts, plays, newspapers, and journals. The sense of the public as a fleeting entity restricted to a specific place and time gradually engendered the broader meaning of a permanent community or group within society. *Das Publikum* was "public" in the sense that admission to it was not limited to the nobility, that is, not based upon the traditional hierarchies of birth and rank, but open to persons, regardless of social origin, whose education and financial means allowed them to participate in this new sphere of activity. It was this public and the cultural and social institutions that gave it a sense of shape and coherence—the print market, public theaters, reading societies, clubs, and coffeehouses—that came to be known as *die bürgerliche Gesellschaft*, or civil society.[19]

The trajectory of the idea of the public in Russia largely paralleled that witnessed in Germany. On one level, the public could denote an audience or body of readers as, for example, in Ivan Krylov's reference to the unusually large "Public" gathered for a performance of the one-act play *The Alchemist* in St. Petersburg on 13 June 1793.[20] Indeed, this meaning has come down to the present day as its primary one, just as it has for the German *Publikum*. The public in Russia during the second half of the century, however, also came to refer to a novel social body formed out of these ephemeral publics of spectators, listeners, and readers. Although most commonly known as the "public" or even the "honorable public," this emerging entity went by other names as well, most notably, "enlightened society" or simply "society."

As was the case in other parts of Europe, the Russian public was largely defined as a social group by its separateness from the masses, the *narod*, or, as they were most often called in the eighteenth century, the *chern'*. This constitutive distinction was explicitly stated in an announcement from the Imperial Academy of Sciences published in the *St. Petersburg News* on 19

May 1780 informing the "honorable public" of the renewed opening of the Kunstkamera (the museum that grew out of Peter the Great's collection of art and curiosities) to visitors at the end of the month. The newspaper instructed its readers to pick up their admission tickets a day before their intended visit in order to minimize overcrowding and thus to ensure a pleasant viewing atmosphere for all. The only groups expressly omitted from this invitation were "[l]ivery servants and the common people" who would not be permitted to enter.[21] Over one hundred years later, the *Explanatory Dictionary of the Living Great Russian Language* recorded the same primary delineation: "Public—society, the people . . . we call society the public, except for the common folk, the simple people."[22]

Of course, it must be recognized that in Russia the distinction between the public and the "simple people" carried an important additional meaning that was lacking in most European countries. Whereas in western Europe the distinction expressed for the most part a *social* division, in Russia it also conveyed an arguably deeper *cultural* division related to Westernization that led to the creation of two distinct worlds—one thoroughly embracing European culture, the other adhering to the traditions of Muscovite Russia. Thus, a particular style of dress or manner of bearing communicated more than one's real or purported social status: it also denoted an explicit and conscious adoption of a foreign cultural model and standard. While this fact is especially important for understanding elite society in the first half of the century, one might argue that it becomes progressively less so in the second half when even those Russian critics who lamented the baleful influence of Westernization did so from within a distinctly European cultural and intellectual framework, such a model having by then been thoroughly internalized.

Whatever the complexities of Westernization, it is important to note that after recognizing this crucial demarcation in the construction of the Russian public, one is still left facing an amorphous social entity whose borders remain poorly defined and difficult to place. Traditionally, the eighteenth-century Russian public has been conceived of as referring almost exclusively to the nobility. Abbott Gleason has recently written that while society (*obshchestvo*) also included some higher-level bureaucrats from prominent families, it continued to denote an aristocratic milieu throughout the century. John LeDonne has argued that even as late as the second half of the nineteenth century "civil society" was equivalent to Russia's "ruling class," which he defined as the "hereditary nobility."[23] Surely, little doubt can be raised about the central position of the Russian nobility within the public before 1800. Still, the two should not be seen as essentially the same. More accurate is Marc Raeff's designation of society as a "'civil society of the educated.'"[24] The crucial distinction here is the highlighting of *education* over *birth* as a primary factor in determining access to this social stratum. Of course, educational opportunities were largely a function of one's birth; nevertheless they cannot be reduced to the exclusive prerogative of a single social group like the nobility, especially in an age that witnessed pronounced expansion of the educational system along with massive dislocations within the nobility itself. As Elise Kimerling Wirtschafter has observed, even at its highest levels the Russian nobility experienced persistent flux in both its membership and its socioeconomic standing throughout the period. The nobility in imperial Russia constituted a "'boundless' social formation touching 'the foot of the throne' at one end and the peasantry at the other."[25] Even Professor LeDonne, who divided Russia into two constituent groups—ruling class and dependent population—admitted that the former represented "a highly unstable world" that was set apart from the mass of Russian society by a large "grey zone," peopled by town and village officials, guild leaders, well-to-do peasants, soldiers' children, and priests' offspring, through which men and women were continually passing in both directions.[26] Although the nobility dominated Russia's public sphere, it shared this space with individuals from various backgrounds and social stations including state officials of different ranks, professors, men of letters, clergymen, noblewomen, merchants, and other representatives of the free professions. What distinguished these Russian men and women from the masses and designated them as members of the public was a shared mode of being that presupposed literacy and a moderate degree of wherewithal, and that was reflected by one's dress, manner of speech, and style of comportment, among other things. The social composition of the Russian public was similar to that in other European countries where the core of the public was formed not from a bourgeoisie in the traditional economic sense, but from the growing ranks of educated state officials and administrators, extending outward to include scholars, military officers, writers, and other social groups that constituted the educated classes.[27]

Disagreements over the composition and definition of the public occupied Russians themselves throughout the second half of the century. In the preface to his tragedy *Dmitrii the Pretender* (1771), A. P. Sumarokov railed against what he perceived as Russian nobles' self-serving attempts to define the public according to notions of rank, wealth, and genealogy, ignoring what he felt were the only valid criteria of discernment and education.

> The word "Public," as Monsieur Voltaire somewhere states, does not refer to all of society, but only to a small portion of it, namely, to persons of taste and learning. If I were to write a dissertation on taste, I would say what it is and explain it. But that is not the point here. As is well known, there are plenty of ignoramuses in Paris just as there are everywhere else, since the large part of the universe is filled with them. The term "common people" [*Chern'*] appertains to the "humble people" [*nizkii narod*] and not to the "base people" [*podloi narod*].

The base people are in fact convicts and other despised creatures and not artisans and farmers. And yet we give this name to all those who are not nobles.
Nobility! What great importance! According to this foolish attitude, the intelli-
gent priest and preacher of God's majesty or, in other words, the theologian;
the natural scientist; the astronomer; the orator; the painter; the sculptor; the
architect; etc.—all of these are members of the common people. Oh, intolerable noble pride, worthy of disdain and desecration! The true common people
are the ignoramuses, even if they possess high ranks, the wealth of Croesus,
and even if their family tree originated with Zeus and Juno, who never existed; with the son of Philip, the conqueror or, rather, wrecker of the universe;
or with Julius Caesar, the securer of Rome's glory or, more accurately, its de-
stroyer. So the word "Public," even where there are many learned persons,
doesn't mean a thing.[28]

Although Sumarokov's characterization of the nobles may have been exaggerated, his depiction accurately conveys that wellborn individuals saw
themselves as members of the public as did those persons of more humble
origin who, through their intellectual and cultural achievements, sought to
distinguish themselves from the common people.

Sumarokov's words are important for yet another reason. They demonstrate that a concern for delineating the precise membership of the public
should not distract from the important realization that, whatever its social
basis, the public figured as an intellectual construct whose borders and exact social referent remained blurred, fluid, and debated. By the second half
of the century, this construct had thoroughly established itself as an integral
feature of educated Russians' mentality, part of the framework through
which they perceived their world. Although the idea of a public was shared
by all, its essential character and function were frequently sources of diver-
gent and mutually exclusive interpretations. On the one hand, the public
might be construed as the locus of a dispassionate and enlightened authority, the final arbiter in settling disputes among society's members, as shown
in the passage from the *Drone* that began this chapter. On the other, the public could be limned as ignorant, lacking in refinement and polish, and consequently, void of taste and critical reason. This was the public Sumarokov
envisaged in his preface, quoted above, when, greatly frustrated with the
staging and reception of his plays in Moscow, he denounced it as crude and
unworthy of his talent and bemoaned having to slave "for the pleasure of
the public."[29] Thus, the status of the public remained ambiguous to the very
men and women who constituted it, who moved within its boundaries and
shared its spaces.

The specific sources of this ambiguity are complex. Some of the confusion can be attributed to the elusiveness and elasticity of the term: lacking a
clearly defined and mutually agreed-upon referent, the public could easily

be invoked for any number of purposes and made to conform to any number of meanings. Such malleability was partly due to the novelty of the notion itself, a symptom of the attempts to assign some stability of reference to an emerging concept. But the varying attitudes toward the public and the apparent existence of "publics" different from one another were also expressions of the social diversity of the body framed by the term. The notion of the public as a coherent and unified whole belied the differences among the individuals who formed it (ranging from powerful aristocrats to skilled artisans), many of whom sought to distance themselves from those they deemed socially inferior.

The public as an intellectual construct was supported by a series of new cultural practices and institutions that together formed the public sphere. These discursive and physical spaces constituted the arena inhabited by the public and helped to define those affiliated with it as members of such an entity. Russia's public sphere operated on two separate levels. First, the print market, or public sphere of print, with its journals, newspapers, and books evolved into a network of communication that reached literate Russians within the two capitals and, by the end of the century, extended far out into the provinces, uniting readers throughout much of the country and linking them to other parts of the world. Second, a host of new physical sites from salons, theaters, and lecture halls to clubs, literary circles, and learned societies presented a zone of social intercourse beyond the traditional spheres of family and state. The appearance of these institutions in St. Petersburg and Moscow as well as in numerous provincial towns marked an increasingly vibrant urban life that became especially pronounced in the reign of Catherine II.[30] The popularity of such venues suggests the degree to which Russia participated, along with the rest of Europe, in what one historian has recently dubbed "the sociable century."[31]

THE PUBLIC SPHERE OF PRINT

The public sphere of print, which first acquired recognizable form during the reign of Catherine the Great, was a product of sweeping changes in the fields of printing and publishing that took place in eighteenth-century Russia. Printing had barely developed in Russia before the reign of Peter the Great. The Muscovite authorities possessed only one publishing operation and the country's few remaining presses were operated by several monasteries located on Russia's western border. All told, these presses produced fewer than five hundred works over the course of the entire seventeenth century.[32] During the next hundred years, and especially between 1750 and 1800, the picture changed radically as the expanding absolutist *Polizeistaat* and then a significant number of private individuals came to recognize the utility and importance of the printed word. Between 1752 and 1774, eight new institutional presses were opened, and by 1770 Russia's Academy of Sciences had acquired approximately seventeen presses of its own, making it one of Europe's largest publishing houses. Catherine's decree of 15 January 1783 on the right to establish private printing presses led to a further expansion of Russia's printing industry. In the final quarter of the century, more than thirty individuals or partnerships opened private presses in the two capitals, and a handful of others rented existing institutional presses. In addition, by 1801 there were twenty-six Russian-language presses active in over twenty provinces.[33]

Russia's greatly expanded publishing base produced an ever increasing stream of works. Between 1755 and 1775, the number of books and journals published yearly rose from approximately fifty to almost two hundred, and by the mid-1780s, four hundred Russian-language books and journals were produced each year. During the five years from 1769 to 1774, sixteen journals were published in St. Petersburg alone. In what is perhaps the most telling statistic of all, eight thousand separate titles were published in the final quarter of the century—over *three* times the number published in the previous *two centuries*.[34] This explosion of printed matter consequently fed the growth of booksellers, whose numbers increased from about fifteen in St. Petersburg and Moscow in the 1770s to well over fifty by the final decade of the century. Russia's provinces boasted some fifty booksellers by 1800 as well.[35]

During this period a truly national print market came into being. Whereas in the first half of the century the public sphere of print was generally limited to the two capitals, during the reign of Catherine II Russia's provinces became an integral part of the book and periodical market. While the flow of printed matter from the country's provincial towns to the two capitals remained small, the provincial reading public began to receive with regularity a steady stream of publications coming out of St. Petersburg and Moscow. The memoirist A. T. Bolotov, for instance, observed how, in the later decades of the century, local society in the town of Bogorodsk began subscribing to all sorts of magazines and journals, whose arrival was awaited with great excitement.[36] Especially popular in Bogorodsk were Novikov's publications. His journal *Morning Light* had more than seven hundred provincial subscribers between 1777 and 1780. In addition, the works Novikov published with the Moscow University Press and the Typographical Company could be purchased in about twenty provincial book shops. A good sense of the geographic range of this market is suggested by the 1786 list of subscribers to the weekly *Mirror of the World,* which included readers in dozens of localities, from Reval in the west to Perm' in the east and Astrakhan in the south.[37]

An increasing diversity in the titles published accompanied the expansion of the print sphere's geographic range. Whereas between 1708 and the beginning of 1725, official pronouncements (for example, laws, manifestoes, and regulations) made up 60 percent of the titles published, by 1787 works in the humanities (history, philosophy, belles-lettres) made up 60 percent, the former category having decreased dramatically to less than 25 percent.[38] These numbers reflected the shift from a period in which publishing and printing were predominantly instruments of state and operated independently of market forces, to a time when publishers largely determined what was being printed. Publishers and booksellers increasingly had to satisfy the demands of a rapidly expanding body of readers characterized by an increasing differentiation of tastes and interests. Gary Marker has noted that, by the 1780s, Russia's "reading public had become large enough to provide an audience for nearly every book for which an audience existed in the West."[39] This development did not obtain complete and unequivocal support, however. Like their counterparts in England, serious-minded publishers such as Novikov and Petr Bogdanovich sought to cultivate the reading public by supplying it with materials they considered to be of social value and were greatly distressed by readers' pronounced appetite for romances and adventure stories, which these Russian publishers were forced to make available in order to remain solvent.[40]

Thus, by the later decades of the century, the public sphere of print had become a dynamic arena that offered the public a wide array of books and periodicals to suit practically all tastes and interests. The diversity of print matter reflected the diversity of the reading public, composed of men and women from a broad range of social levels, occupations, and education. As the poet I. I. Dmitriev remarked in 1789, not only did "educated people" read, but so too did "merchants, soldiers, serfs and even gingerbread and bread salesmen."[41] Yet just as the print sphere mirrored the diversity of Russia's readers, it also served to produce among them a sense of shared or collective identity as a direct result of their access to and participation in this sphere. In an age of massive illiteracy, the ability to read marked one as unique and separate from the majority. The print sphere made literate individuals more cognizant of each other by linking them in a network of shared communication. As participants in this network, Russian readers came to see themselves as part of a larger collective, that is, the "honorable public" so frequently addressed in the day's journals and newspapers. As W. Gareth Jones has shown, this new sense of shared identity was not simply an unintended result born of the common consumption of print. Journals like Novikov's *Drone* explicitly sought to provide its body of readers "a self-image, to give it a sense of corporate identity." The journal's readers, he added, were made to feel as if they were members of the same club, one whose meeting place was the pages of the *Drone*.[42]

But the Russian public had many more meeting places at its disposal than the pages of journals such as the *Drone*. For along with this discursive sphere, there arose a number of new physical spaces that, like the print market, brought together Russia's educated classes in a novel way and formed the basis of Russian civil society.

SOCIAL SITES OF THE PUBLIC SPHERE

Late in 1718, Peter the Great issued an ukase—drafted by the St. Petersburg head of police Devier (but edited by the tsar himself)—creating a strange new institution called *assamblei*, that is, assemblies or evening parties. According to this ukase:

> Assembly is a French term that cannot be conveyed by a single word in Russian. It may be said to denote an informal meeting or gathering in a home intended not only for amusement, but also for affairs; for here one can see others, discuss all matters of importance, hear the day's events, and, moreover, enjoy oneself.[43]

Held in private homes in the new capital two or three times a week during the winter months, the assemblies—with their dancing, games, drinking and eating, and polite conversation—were intended to create a heretofore unknown opportunity for relaxation and agreeable amusement. The ukase stipulated that all persons of high rank as well as military officers, prominent merchants, craftsmen (such as shipbuilders), and state officials, in addition to their wives and children, were permitted, and largely expected, to attend.[44] In the residences of Chancellor F. A. Golovin and Prince A. D. Menshikov, for example, these men and women gathered to enjoy each other's company and to take part in a form of social interaction that was to be "informal," meaning the participants should come and go, sit, move about, and talk with whomever they wished in a relaxed, informal, and "natural" manner.

Peter's establishing of the assemblies represented a radical innovation in the lives of Russia's elite and provides an excellent point of entry for an examination of the host of new venues and institutions of the public sphere that developed during the century. Although social gatherings in the homes of the wealthy and powerful had not been completely unknown in Muscovite Russia (those held at V. V. Golitsyn's Moscow residence in the latter part of the seventeenth century, for example, are best remembered)[45] the novelty of the assemblies cannot be denied—a fact that is vividly conveyed by the need to introduce them with an official edict that described in detail not only what precisely they were, but even how they were to be conducted, who was to take part in them, and how one was expected to behave there.[46] The assemblies marked a decisive departure from the norms of the Muscovite elite in several fundamental ways. First, their sociability denoted a shift from the pronounced monasticism that characterized the lives of Muscovy's ruling classes. Earlier, social gatherings had been limited to significant moments in the life of the family or of the larger community of Orthodoxy, acknowledged through feasts of varying proportions. The form of socializing prescribed in Peter's ukase, with its secular music, dancing, and free social mixing, was largely unknown in Muscovite Russia. The situating of the assemblies "in private homes" represented a second change. The tsars of Muscovy generally did not visit others; rather, they let the world come to them. Peter's officially designating private homes as the proper venue for the assemblies and his wholehearted participation in these events lent to these sites a new legitimacy and significance that they had not formerly possessed.[47] While the court remained the primary goal toward which the Russian elite directed its energies, new centers of cultural and social activity—such as these private residences—arose in ever greater number and grew in significance over the course of the century. Finally, the assemblies "assembled" groups that traditionally had limited social intercourse, bringing them together in a decidedly original manner. In the past a boyar would have resisted inviting a master craftsman or merchant as a guest, for to do so was to risk a dangerous loss of honor. The interaction of such groups was limited to special occasions that were governed by an elaborate ritual code meant to reflect in concrete form the inequality of the participants and their relative position within the social hierarchy.[48] But in the assemblies, groups of unequal rank and standing mingled on a more or less equal footing: the traditional rigid ceremony that marked social boundaries and hierarchies in daily interaction (and especially in the receiving of guests) was quite explicitly rejected.

The inclusion of women required setting aside one of the chief divisions within Muscovite society, that of the inequality of the sexes, and ushered in a new era in the lives of Russian females. Before the reforms of Peter the Great, women in Russia's upper classes led lives of extreme social isolation including mandatory seclusion unlike anything known in the West. Largely confined to their own quarters *(terem)*, they inhabited a world apart from men, with whom they were not allowed to mix socially. On those occasions when the boyars' and princes' daughters and wives did leave their quarters, usually only to attend church services or to visit kinsmen, they were hidden behind curtains or moved about in closed carriages. Almost invisible in the larger social world, Muscovite noblewomen, to quote one leading specialist, "were studiously excluded from public life" and were subjected to a "rigid . . . regime of social control." Whatever his ultimate motives in introducing the assemblies, perhaps a desire both to end Muscovite politics based on marriage alliance and to introduce western European cultural forms, Peter the Great quite literally freed Russia's noblewomen from the extreme patriarchy of Muscovy and brought them out of the private realm of the *terem* and into the public sphere of society. This fact attests to women's inclusion in Russian civil society from the eighteenth century and disproves the notion that this social formation was "gendered" to exclude women. The birth of educated society in Russia did not mark the forced removal of women from the public (or, male) to the private (or, female) sphere; rather, it signified the exact opposite. Peter created with the assemblies a model for educated society that recognized the participation of both sexes. It was a model that, despite some exceptions, retained its power throughout the entire century.[49]

Although implemented from above by tsarist decree, the assemblies—and the form of sociability practiced there—were not simply imposed by force onto an unsympathetic and recalcitrant society. Rather, these cultural imports landed on receptive soil, quickly becoming constituent features of the lives of Russia's privileged classes. By the second half of the century, the large homes and palaces of the notables in and around St. Petersburg had become centers of a vibrant social life. Many of the public's leading figures maintained what can be called "open" houses that hosted an unending stream of balls, masquerades, banquets, and concerts. There were between fifteen and twenty salons in the capital.[50] Important as these activities were in the lives of the nobility, other groups shared the pleasures as well. Along with the high government officials and prominent aristocrats who frequented the residences of L. A. Naryshkin and A. S. Stroganov, for example, were actresses and actors, artists, musicians, and *gens de lettres*.[51]

One of the liveliest homes in St. Petersburg belonged to I. I. Shuvalov, who maintained a literary salon popular with figures such as Sumarokov, M. V. Lomonosov, G. R. Derzhavin, and I. F. Bogdanovich (the author of *Dushen'ka*). Here these leading literati engaged in endless discussions and debates, best known of which were the fiery (and occasionally drunken) disagreements between the uncompromising opponents Sumarokov and Lomonosov. The publishers V. V. Tuzov (of the journal *Day-Labor*) and V. G. Ruban (of *Neither This Nor That*) used these gatherings as a chance to continue the debates then raging in the periodical literature and to seek both intellectual supporters as well as subscribers. But Shuvalov played host to more than St. Petersburg's literary community, and over the years his other guests included the empresses Elizabeth and Catherine II; the emperor Peter III; Princess Dashkova, playwright, president of the Russian Academy of Sciences, and director of the Russian Academy; N. I. Perepechin, the famous art patron and a director of the Assignat Bank; and Kirillo Kamenetskii, Shuvalov's personal physician and the author of a popular herbal. Along with their various literary discussions, Shuvalov's guests passed the time playing cards and discussing the latest local and foreign news.[52]

A second important literary salon met in the early 1770s in the home of the poet and novelist M. M. Kheraskov and his wife, the poetess Elizaveta

Kheraskova. The Kheraskovs' salon attracted numerous belletrists and other "lovers of philology" such as Bogdanovich; the poets V. I. Maikov and A. A. Rzhevskii; A. V. Khrapovitskii, Catherine II's future secretary; and his sister, M. V. Khrapovitskaia-Sushkova. They presented their literary creations and translations, many of which were then published in the circle's own weekly, *Evenings*. It was in *Evenings* that the first Russian translation of Edward Young's *The Complaint, or Night-Thoughts on Life, Death, and Immortality* (1742–1746), attributed to Khrapovitskaia-Sushkova, appeared, representing one of the earliest expressions of pre-Romanticism in Russian literature.[53]

Other popular venues were the residences of Chancellor A. A. Bezborodko, who regularly entertained distinguished foreigners, high state officials, and leading figures from the spheres of art and science in his palace on Novo-Isaakievskaia Street, and of S. V. Saltykov on Malaia Morskaia Street, which was the site for "les mardis européens"—dance parties frequented by the capital's fashionable society every Tuesday evening. On the same street a more egalitarian sociability reigned in the home of the retired general A. I. Arbenev, where in the 1780s the highest ranks of Russian society mixed with their social inferiors. As the memoirist I. I. Vigel later described it: "[N]otables said that they went to the home of Arbenev to laugh at the society gathered there, but if one were to tell the truth, they went to enjoy themselves . . . young people came every evening, the house was packed, everyone laughed, everyone danced; true, they only burned tallow candles and kvass was provided as refreshment; indeed one didn't ask for any fancier fare, but what merriment, the liveliest merriment, which is truly better than the mere luxury, which has replaced it in our time."[54]

The great love of dance exhibited at these soirées began in the 1730s and quickly spread to include members of different social stations. The organizers of dance balls soon included everyone from Russian military officers and foreigners of various ranks to tailors and valets who regularly rented out space in the homes of wealthy nobles and merchants, hired musicians, and sold admittance tickets. Certain balls were attended mostly by guards officers, others by a mixture of officers, merchants, skippers, and petty officials; those dances thrown by valets attracted primarily "'lackeys and similar types.'" While officers' balls tended to be rather exclusive, admitting only those personally invited, balls put on by men of lesser social standing were open to all those who heard about them by word of mouth. In order to attract as many women as possible, only men had to pay the standard admission price—usually between fifty kopecks and one ruble, depending on the social status of those in attendance.[55] By the end of the century, balls were often divided into two types: "nobles" balls, extremely exclusive affairs limited to a select group of the capital's notables, and "English" balls, more open gatherings made up of nobles and merchants.[56]

Public concerts also found a home in the palaces of St. Petersburg. During the middle of the century, Count A. G. Razumovskii, thought to be the secret husband of Empress Elizabeth, put on concerts as well as balls and masquerades in a large pavilion alongside the Anichkov palace, which housed his extensive painting collection. In the 1790s, performers and artists organized public concerts and masquerades in this same pavilion, which had earlier belonged to Prince Grigorii Potemkin. The prince had often hosted such events. In the winter of 1780, an advertisement in the *St. Petersburg News* informed readers that in the home of Count Osterman a Mr. Lolli was presently selling tickets (five rubles for all three concerts) for an upcoming concert series in Potemkin's residence.[57]

Equally popular were the numerous gardens and parks in and around the city, for they provided a pleasant alternative setting for the socializing that during most of the year was confined to the indoors. The imperial gardens—for example, Tsaritsyn Meadow (later known as Mars Field), the Summer Garden, and the gardens at Ekaterinhof—were the most widely used, serving as locations for various outdoor parties and festivities. At such outings the members of the public sought to ensure that their contact with the common folk, who also flocked to such events, was kept to a minimum, that the social distance that set them apart was given an equivalent physical expression. Thus, before an upcoming fireworks display on the Tsaritsyn Meadow in 1780, for example, a Mr. Kinetti notified the "most honorable public" that seats in his small, yellow seating gallery were on sale for one ruble a piece in Mr. Marsani's coffeehouse in the Summer Garden.[58] While both the public and the masses shared in the excitement of the fireworks, they did so from vantage points that were separate and that reflected their relative position in society.

Many of the notables renowned for hosting winter soirées, such as Stroganov and Bezborodko, opened their private gardens during the summer months, and the public gathered there on holidays for outdoor fêtes. One of these "pleasure gardens" on Round Island was open on a paid basis—twenty kopecks per person for a single visit or two rubles fifty kopecks for the entire summer. An advertisement in *Idle Time for Good Use* in June 1759 announced the opening of the garden at the Corps of Cadets each Thursday and Sunday to "persons of any rank and title, except for nobles' servants in livery. . . as well as those persons dressed in foul attire." The garden's numerous attractions included a summerhouse where one could find newspapers (from Russia, Germany, France, and the Netherlands), billiards and other amusements, and food of all kinds: lemonade and coffee, ice cream and candy, and, for groups of six or more, dinner especially prepared by Otto Luko, the school's own chef.[59] In the spring of 1793, Baron Ernest Wanžura, a court pianist and composer who took an active role in managing the imperial theaters, opened the "Vauxhall in Naryshkin Garden" on the Moika Canal, what one historian of the city has called the first public pleasure garden. The vauxhall hosted all sorts of festivities, dances, and masquerades every Wednesday and Sunday evening. For a one-ruble entrance fee visitors danced to one of the garden's two orchestras, watched a stage performance in the open theater, or simply marveled at the assortment of human and animal curiosities on display ranging from strong men and "giants" to lions and other rare beasts.[60]

In addition to Baron Wanžura's vauxhall, St. Petersburg had other venues for enjoying dramatic performances, and over the course of the century the city witnessed the birth of a lively theater life. European theater culture had not been unknown in pre-Petrine Russia, although it was largely restricted to the court and the private theaters of several leading boyars. Peter sought to establish Russia's first true public theater when in 1702 he invited the well-known actor Johann Kunst and his wife to leave Danzig for Moscow, along with seven actors. Kunst died the next year before he could put on any productions for city residents in the newly constructed theater on Red Square, but his place was quickly taken by Otto Fürst, a local German. After some initial success, Fürst's theater proved a failure, and in 1706, after having been moved to the Kremlin, it was closed. The remainder of Fürst's theater gained new life, however, when it was transferred to the village of Preobrazhenskoe near Moscow to form the basis of Natal'ia Alekseevna's personal theater. The younger and much beloved sister of the tsar, Natal'ia Alekseevna staged a variety of religious, didactic, and comedic plays for a select public at Preobrazhenskoe for several years before moving to the new capital, where she carried on with the productions until her death in 1716.[61] Another important source for the development of theater culture in Russia was the so-called school theaters located in seminaries and academies. Moscow's Slavonic-Greek-Latin academy began staging school plays around 1700, and by 1750 school theaters were active in Rostov, Astrakhan, Novgorod, Tver', Tobol'sk, and St. Petersburg, where Feofan Prokopovich, the leading church figure in Peter's reign, opened a school theater in the early 1720s. These dramatic productions served a variety of purposes, most notably propagating the achievements of Peter and civilizing and enlightening both the young pupil-actors and their small audiences.[62] The Corps of Cadets (founded in 1731), whose students established their own dramatic company, which performed at the school as well as at the Opera House and at court, was of particular importance for the development of theater in the new capital. The young actors' productions of Racine, Molière, Shakespeare, and even of one of their fellow pupils, Sumarokov, who also acted as the group's director, were generally well received by the St. Petersburg public, and the company continued to operate into the late 1750s.[63]

Although companies like the one attached to the Corps of Cadets usually played to very select audiences, by the middle of the eighteenth century

European theater was no longer the preserve of the court elite and a hand-ful of grandees. During the reign of Empress Elizabeth (1741–1761), "emi-nent tradesmen" and soon thereafter merchants and their wives began to frequent the Summer Garden's Opera House that had formerly admitted only the most privileged members of Russian society.[64] On 30 August 1756, Elizabeth issued a decree creating Russia's first permanent professional the-ater open to the paying public. Headed by Sumarokov and initially located in Count Golovkin's stone house on Vasil'evskii Island, the new theater soon moved to the more convenient Opera House where tickets sold for one and two rubles. In addition, a privately run German theater had opened its doors several years earlier (around 1752) on Bol'shaia Morskaia Street.[65] The trend toward opening up the theater to a broader spectrum of the capi-tal's inhabitants continued under Catherine II. This trend was evident in the construction in 1783 of the Large Stone Theater on the city's outskirts, an unprecedented venue intended for lavish productions and meant to seat au-diences of up to two thousand. Karl Kniper, a German impresario who had earlier been active in Moscow's theatrical life, ran St. Petersburg's other ma-jor public theater, most commonly referred to as the "city" or "wooden" theater to distinguish it from the Large Stone Theater. Kniper's playhouse had its heyday in the 1770s and 1780s when it staged a variety of comedies, comic operas, and small ballets.[66]

By the final quarter of the century, the theaters of St. Petersburg had be-come truly popular institutions to which the city's residents flocked in in-creasing numbers. Whereas in 1754, Elizabeth was so disappointed by the low turnout for a French comedy at the Opera House that she obliged the nobility and members of the upper classes to attend future productions or risk being fined, by the late 1760s the theater-going public had grown to a size that made it a subject of satire for journals such as *All Sorts of Things*.[67] Although the purported reason for attending the theater was to enjoy the words and actions presented on stage, a good many in the audience pre-ferred to direct their attention to their neighbors, using the occasion to chat-ter with friends and acquaintances and to catch up on the latest gossip, all of which lent the theater the air of a social club or coffeehouse. In the 1770s, the weekly *Evenings*, the organ of the Kheraskovs' salon, published a letter from one irate woman who complained that during a recent visit to the the-ater most of those attending had come solely to show themselves off and to delight in the sight of their fellow spectators. Constantly prattling and dis-turbing the few who sought to view the scheduled performance, such per-sons, she went on, had no idea of how to behave properly "in gatherings" or in the presence of "society."[68]

Others shared this distaste for the unrefined types making their way in ever increasing numbers to the theater and turning it into a site of great so-cial mixing.[69] One response to the situation was to segregate the audience according to rank and station by reserving the loges and boxes for the "better sort" or even, as in the Large Stone Theater, by constructing a special entrance leading directly to the cheap seats in order to protect the capital's finest from any unpleasant contact with their social inferiors.[70] But rank was not always a guarantee of refinement. In her letter to *Evenings,* the irate patroness singled out for special criticism the crudeness displayed by Russian nobles—a particularly sad fact, she reasoned, since they should be well versed in the principles of proper deportment. That even nobles could not be counted on to act appropriately in public theaters, that refinement or lack thereof was not solely a correlate of social rank, fed yet another response, that is, the growth of intimate, exclusive theaters in the palaces and on the estates of the court, the aristocracy, and the wealthiest of the nobility. In these places, such as the Hermitage Theater, and in the numerous serf theaters, the highest ranks of society could withdraw from the larger public into a more private and exclusive realm.[71]

Individuals seeking diversions of a more intellectual sort than might be found in the capital's theaters, pleasure gardens, and dance halls could turn to the Academy of Sciences' public assemblies open to "all lovers of the useful sciences." Held in the house of the diplomat P. P. Shafirov on 27 December 1725, the first assembly—which opened with a special ceremony and whose proceedings were duly recorded by an attending protocolist— attracted close to four hundred persons. They came to hear speeches by the physicist and philosopher G. B. Bülfinger on the creation of the academy and by the mathematician Ia. German on advances mathematicians had made in the study of longitude. Delivered in Latin, these inaugural lectures reached only a very limited circle; yet by 1750 speeches were given in Russian as well, and translations of the Latin presentations were handed out before each assembly, thereby ensuring a larger audience. As the century progressed, the academy's members drew on an ever wider range of topics in the hope of attracting a broader segment of the public. Following the Lisbon earthquake of 1753, for instance, Lomonosov gave a lecture entitled "A Speech on the Birth of Metals from the Trembling of the Earth," and several years later S. G. Domashnev, the academy's director under Catherine the Great, held a discourse on why their age was commonly known as the "philosophical century." In addition to the public assemblies, the academy's members instituted a weekly lecture series open to the public and advertised through special printed announcements.[72]

Members of the public also frequented the academy's Kunstkamera and library, whose joint opening on Vasil'evskii Island was announced by the *St. Petersburg News* in the fall of 1728. From its earliest days, the Kunstkamera was especially popular, so much so that in the 1770s tickets had to be introduced to keep the crowds from getting too large. The Kunstkamera's wondrous and exotic oddities—such as the life-size wax effigy of Peter the

Great, the "'dead body of an infant with 6 fingers,'" and an "'oddly shaped chicken egg'" from Solikamsk in the Perm' province, to give only a brief sampling—attracted persons of diverse standing who by the end of the century ranged from those of the beau monde, to petty officials, merchants, and students.[73] The academy's library proved quite successful with the capital's educated classes, and they sought to avail themselves of its resources in ever greater numbers. Originally open for only two hours twice a week and used primarily by those affiliated with the academy, after 1752 the library extended its operations to four hours daily every weekday in order to serve the increasing numbers of curious readers, such as courtiers and notables, men of letters, educators, officers, and the academy's students, translators, and academicians.[74]

Along with the Academy of Sciences library, several institutions catering to the public's appetite for printed matter also appeared. Beginning in the 1780s, for instance, the Corps of Cadets opened its library to "all scholars and lovers of science" (except those "in vile clothing"), and in 1756 the Free Economic Society organized a library that also offered the public frequent exhibitions and lectures.[75] Especially noteworthy was the creation of several reading societies and so-called reading cabinets that received strong support from the clergy, merchants, and members of the lower-middle class (*meshchane*). The German Reading Society, founded by the academician A. I. Gil'denshtet in 1777 and attracting sixty members by the mid-1790s, was the first of these societies. Members of this organization each paid annual dues of ten rubles that went toward the purchase of books. After circulating among the society's readers, the books were sold and the proceeds used to fund future acquisitions. Gil'denshtet's society soon spawned others. In 1780 a short-lived French Reading Society was set up, and in 1793 a librarian by the name of Busse established the German and French Reading Society, which quickly grew to over forty members. Despite their names, these societies had Russian members, and Russian-language books were available along with works in German and French. The success of the new societies prompted the city's booksellers to open paying reading cabinets at their stores or, as they were then often advertised in the local press, "public Libraries," six of which operated in the later decades of the century. While western Europeans ran most of these, Russian booksellers participated in this development as well. Perhaps best known among the Russians was M. Ovchinnikov who between 1784 and 1799 operated a paying library called the "Russian Reading Establishment" in his St. Petersburg bookstore.[76]

. . .

*E*ven if on a somewhat smaller scale, society in Moscow developed along similar lines, and by the end of the century Russia's other capital also boasted a vibrant public life.[77] Just as Peter the Great's assemblies were gaining acceptance in St. Petersburg, Feodosii Ianovskii, then acting head of the Holy Synod, introduced them in Moscow in the 1720s among his fellow clergymen. Over the years the homes of the city's wealthy and influential residents came to serve as the setting for a range of social gatherings.[78] The palatial residences of the Golovkins, the Chernyshevs, and Vladimir Orlov, among others, along Nikitskaia Street and its adjoining lanes and alleys witnessed an unending series of concerts and balls during the autumn and winter months.[79] Indeed, after 1750, Moscow, where balls and masquerades figured more prominently in the cultural life of the city than was the case in St. Petersburg, hosted up to fifty such events a month.[80] Having returned to Moscow around 1775, Kheraskov and his wife revived the salon they had maintained in St. Petersburg, and their new salon became the setting for lively gatherings of the city's literary world.[81] Around the same time Prince M. M. Shcherbatov hosted his own salon, and the homes of the brothers Iurii and Nikolai Trubetskoi also became popular meeting places for men of letters, artists, travelers, and others from polite society.[82] While participation in many of these levees at private homes required a personal invitation or at least an acquaintance with those in attendance, this was not always necessary. In the spring of 1756, for example, the widow of General Litskin advertised in the *Moscow News* tickets (one ruble a piece) for a weekly concert series in her home in the German Quarter.[83]

In addition to balls, salons, and concerts, these sites offered other activities, some of which, from a present-day vantage point, appear rather odd or unexpected. Prince Iu. V. Dolgorukov's residence on Tverskaia Street, for instance, served as the locale for a traveling zoo brought to the capital by the "Italian Antonio Belli and company." For twenty-five kopecks (the *Moscow News* informed its readers), the curious could view these "diverse beasts from foreign lands"—for example, "a leopard of the greatest size and beauty, a mandrill with a blue face . . . various birds and a stuffed lion"—all safely locked in strong metal cages.[84] For those with slightly different tastes, in 1762 the merchant Telepnev put on a cryptic-sounding show in his home near the Nikitskii Monastery that featured a "young man with no legs and his art with all manner of astonishing things."[85]

Especially popular with the Moscow public were the tsars' country palaces (for example, Petrovskii, Tsaritsyn, and Izmailovo) and the estates of the nobles whose "gardens and establishments," as one eighteenth-century observer noted, "are open to the public."[86] Among these estates, the Sheremetev family's Kuskovo, the setting for countless festivities and celebrations, lavish feasts, spectacular fireworks, and elaborate dramatic performances, enjoyed a unique status. Filled with amusements and exotic structures intended to entertain and delight, Kuskovo's expansive park and gardens presented the Sheremetevs' guests with an enchanting world to explore. Here they might pause during their strolls to marvel at one of the curious houses constructed in the architectural styles of China, Holland, or Italy or to savor a moment of quiet reflection at the Temple of Silence or the small Philosopher's House, where the sign over the door instructed all who entered to "Find Tranquillity Here." For those seeking more active entertainment, there was a vauxhall offering music and dance. Finally, Kuskovo's large open-air theater staged elaborate performances every Thursday and Sunday during the summer.

The celebrations at Kuskovo maintained some of the spirit of the more traditional fêtes *(gulian'ia)*. Mass outings of up to fifty thousand people, they attracted even the lowest ranks of the common folk, but members of Moscow society did not blend in with these common people: their ranks were demarcated by the fact that they, unlike the masses, received written invitations to the festivities at Kuskovo and sought different recreations there. While undoubtedly all in attendance could enjoy the fireworks, an appreciation for and the privilege of entering the Temple of Silence or the Philosopher's House marked one as a part of the Moscow public.[87]

Wealthy nobles, however, were not the only ones to provide such entertainment. In the late 1760s, the entrepreneur Melchior Groti opened Moscow's first vauxhall in a garden near the Donskoi Monastery, which he rented from Prince P. N. Trubetskoi. Several years later, following the outbreak of plague, Groti, together with Prince Urusov, switched locations and began leasing the country home of Count Saltykov not far from the Exaltation of the Cross Church, where throughout the summer they entertained a largely noble gathering with dances and dramatics twice a week.[88] Following Groti's departure, the Englishman Michael Maddox took over the operation of Moscow's vauxhall.[89] In 1783, Maddox purchased a large piece of land in the Taganskaia area from the merchant Savva Iakovlev and built there a brand new vauxhall. Maddox's establishment offered visitors the amusements found in similar venues: light operettas and one-act comedies performed on the summer stage, followed by a ball or masquerade, and usually concluding with a large dinner, unless a special occasion required a fireworks display. The grounds contained pleasant gardens, a gallery for walks (illuminated at night), and a separate billiard room. According to M. I. Pyliaev, the "English Vauxhall" became quite popular by the 1790s, frequently attracting thousands of patrons each paying a one-ruble entrance fee, an additional four rubles paid by those staying for dinner.[90]

Maddox was also the proprietor of Moscow's largest public theater, the Petrovskii, built in 1780.[91] Unlike existing theaters that typically held an audience of no more than a few hundred, the Petrovskii Theater was an immense structure able to accommodate two thousand spectators.[92] Ticket prices ranged from one to one thousand rubles—the yearlong subscription price for the most expensive boxes. According to one of the theater's visitors, the Englishman William Tooke, the Petrovskii was "generally full" and the loges were rarely "unlet" despite the rather high price of admission.[93] For some, however, theaters such as the Petrovskii were becoming too popular, too accessible, admitting all those able to pay the minimum admission price. As in St. Petersburg, the highest strata of society in Moscow increasingly preferred the entertainments available to a select few in the more private setting of aristocratic palaces where one was assured of being safe from any unpleasant contact with the common sort. To attend public theaters was less "refined," as one contemporary remarked, than to attend performances "by the personal invitation of a host, and not where anyone could be present in exchange for money. And who, indeed, among our intimate friends did not possess his own private theater?"[94] The retreat to private residences for theater and concert performances closed to the hoi polloi was not unique to Russia but was also evident in England around the same time and reflected a more general European trend.[95]

Along with its regular dramatic performances, the Petrovskii hosted concerts and masquerades in separate halls built specifically for these events. Masquerades, admission to which cost one ruble, were held in the rotunda: a spectacular circular hall able to hold thousands, lined with mirrors and illuminated by more than forty crystal chandeliers. Concerts, frequently given during Lent, ranged in price from two to five rubles, and they too never failed to attract large audiences. Finally, the Petrovskii Theater even served as the venue for public lectures, such as the one given by the physicist Filidor who demonstrated his various experiments before an audience in the theater's concert hall.[96]

The primary setting for such learned presentations was Moscow University where, since its founding in 1755, public lectures and disputations played a prominent role. Open free of charge to all "lovers of science," these lectures, as one scholar has noted, "nourished and maintained the constant attention of the Moscow public of all categories" and helped to integrate the university into the intellectual and social life of the city.[97] The public presentations came in two basic forms—the "public speech-days" and the "public courses." The former, advertised in advance by special placards printed in Latin and Russian posted throughout Moscow, were noteworthy events full of pomp and solemnity. Held in the university's assembly hall, speech-days consisted of presentations and disputations by and among university and high school students and became significant social occasions attracting "the higher clergy, distinguished personages of Moscow, foreigners, and all of educated society."[98] The university published the speeches in Russian, copies of which were handed out immediately following the presentation.[99] As well as being fashionable social events, a chance to see and be seen by important figures of Moscow's polite society, speech-days were also a significant institution in the intellectual life of eighteenth-century Russia. Along with the growing public sphere of print, they functioned as a major forum for the dissemination and discussion of a wide range of new ideas. It was not uncommon for speech-days to be the setting for significant debates, perhaps best illustrated by the scandal that erupted when Archbishop Ambrose of Moscow reacted with hostility to D. S. Anichkov's public defense of his dissertation "Concerning the Origin and Occurrence of the Natural Worship of God."[100]

The second type of public presentation, the "public courses," were essentially regular lectures given daily, Saturdays and Sundays excepted, by university professors. Announcements in the *Moscow News* and special catalogues listing the university's courses notified the public of upcoming lectures. One of the most popular of the university's lecturers was Johann Georg Schwarz, professor of philosophy and a prominent personality among the Masons in Moscow. His lectures on aesthetics (delivered in Russian) attracted large crowds and helped to create for himself a loyal group of devotees drawn from the ranks of university students and the city's wealthy and influential residents. Every Sunday in his home, Schwarz treated his most favored followers to exclusive, private lectures on his theories of knowledge and history, until rumors about these sessions—which had spread throughout the city creating intense curiosity among those excluded—eventually forced him to deliver the lectures publicly at the university. Schwarz's popularity did little to endear him to his academic colleagues who came to resent the German's greater success at drawing an audience and hence at acquiring useful patrons.[101]

Along with the assembly hall and lecture rooms, the public also had access to the library that, soon after the university's founding, opened to all "lovers of Science and enthusiasts for reading books" for three hours every Wednesday and Saturday.[102] In time more libraries were established in Moscow. Novikov operated a successful library–reading room in his bookstore, and beginning in 1778 the city's several French-language bookstores also opened their own reading libraries.[103]

. . .

While the two capitals remained the key centers of the public sphere, the final decades of the century witnessed the rapid extension of this sphere to the provinces as well.[104] The growth of the book and periodical market played an important role in linking literate Russians to networks of communication throughout the country. A crucial factor feeding this development was the emergence of a provincial readership, a direct result of Peter III's abolition of obligatory noble service and Catherine's administrative reforms of the 1770s and 1780s. In the wake of these transformations, rising numbers of educated Russians took up residence outside the capitals, and consequently, a hitherto unknown public life modeled on that of St. Petersburg and Moscow developed in many of Russia's provincial towns.

The establishment of new provincial administrative institutions under Catherine served as a particularly strong stimulus to the enlargement of the public sphere. The reforms of the local administration brought a dramatic upsurge in the number of officials inhabiting the provinces. In 1774, a year before the reforms began, approximately 12,000 state officials manned the local administration, by 1781 the number had risen to 22,000, and by the end of the century it had reached almost 30,000.[105] These officials introduced the same cultural practices and institutions that dominated elite life in the two capitals. The establishment of each new province was marked with a round of lavish festivities: balls, masquerades, musical and operatic performances, poetry readings, and fireworks. The social gatherings did not end with these inaugurations, however, and the governors' residences quickly became central venues in the cultural life of the entire local society. Every three years the governor summoned the assemblies of the nobility, created throughout the countryside by Catherine's reforms of 1785, which often met in a large hall in his residence. The assemblies usually took place in December or January during the height of the social season, and added to the string of various balls, dinners, and fireworks.[106] For the young, the governor's house was a place to play games, dance, and court; for those of more advanced age, it was a place to discuss literature, art, and even politics with one's peers. Some governors possessed large book collections that functioned as quasi–public libraries open to individuals from a range of social stations with shared intellectual pursuits. The home of the poet G. R. Derzhavin was just one example of this new provincial life. While governor of Tambov in the 1780s, Derzhavin hosted assemblies and small dances every Sunday evening, concerts every Thursday, and various theatrical productions on special occasions. While the governor's residence served as the primary locus of society, the homes of wealthier nobles came to provide another setting for the Russian public in the provinces.[107]

As in the capitals, theater occupied a prominent position in the cultural life of the local elite, and the homes of provincial governors served as the setting for frequent theatrical performances, the actors often being drawn from the ranks of the local nobility and civil servants. In 1787 in Voronezh, the governor-general, V. A. Chertkov, created a theater under the direction of Prince Ukhtomskii who, along with the governor's children and local nobles, put on various comedies and tragedies. Theaters soon began to function more independently of the governors when separate theater buildings were constructed and professional acting troupes directed by impresarios replaced amateur thespians in towns like Tambov, Vologda, Voronezh, Kaluga, and Khar'kov.[108] In other towns the process had begun even earlier. By 1750, F. G. Volkov was operating a public theater in Iaroslavl' that had its own building and was funded by ticket sales. In 1760, the writer and dramatist M. Verevkin opened a public theater in Kazan' that despite seating four hundred was often packed to overflowing.[109]

While the provinces lacked the academy of St. Petersburg or Moscow's university, people there did not always feel the absence of the learned amusements that these institutions provided. In Russia's smaller towns and villages, Orthodox seminaries played an important role by putting on annual or biannual theological debates that attracted the local nobility and the rest of elite society. Intended as occasions for young theological students to demonstrate their religious training, the debates became lively social events that (along with the theological discussions) included poetry readings, musical performances, and refreshments.[110] Finally, in the second half of the century public libraries and reading rooms attached to local bookstores were established in several provincial towns including Tula, Kaluga, and Irkutsk, the last boasting Russia's largest public library, founded in 1782 by volunteer contributions and open free of charge.[111]

CLUBS, CIRCLES, AND SOCIETIES

In addition to the new practices and social settings that defined the Russian public and gave it structure and coherence, such as the expanding market of print, salons, vauxhalls and theaters, lecture halls and libraries, there arose during the century equally important clubs, circles, and societies. Founded for a variety of purposes, these new sodalities brought together Russia's educated classes in novel ways, forging bonds of community and affiliation among them and constituting another important thread in the fabric of civil society.

The first club in Russia may well have been the Neptune Society, which met in the early years of the eighteenth century in Moscow's Sukharev Turret, home of the new School of Navigation and Mathematics. Little is known about this "secret club." Its members included Tsar Peter, who served as the society's supervisor; Francis Lefort, its chairman; Feofan Prokopovich, its orator; the Scotsman Henry Farquharson, a mathematician and astronomer who came to Russia to help set up the School of Navigation and Mathematics; and Prince A. D. Menshikov. What precisely transpired at their nocturnal gatherings is not clear, although it is believed their energies centered on conducting certain scientific experiments, similar to those taking place in the clubs of virtuosi then popular in England.[112] The example of the Neptune Society did not catch on, and clubs did not come into fashion until the final decades of the century.[113] One of the first clubs, founded in the early 1770s by a German merchant residing in the capital, was the Large Bürger Club, perhaps better known as the Schuster Club in honor of its founder. The Schuster Club enjoyed great popularity, attracting state officials, wealthy Russian and foreign merchants, prosperous craftsmen, and various artists. Although established as a social club, charity also figured in its range of activities: a form of pension was set up to support 150 aged and indigent persons, and several orphans were raised on funds provided by the club.[114] Around 1790, P. I. Melissino formed the Philadelphia Society, which attracted young men from the capital. The precise aims of the society are not clear, although one observer claimed that its "libertine" members were chiefly interested in "giving themselves over to debaucheries."[115] Other clubs active in St. Petersburg included the American Club, founded in 1783 as an outgrowth of the Bürger Club, which counted six hundred members by 1800, and a series of dance and music clubs. The most popular dance club, founded in 1785 by the master craftsman Ulenglug and open only to "unranked [nechinovnyia] individuals belonging to the commercial and merchant estates," survived into the second half of the nineteenth century. A second "commercial-class dance society" opened in January 1790. The city's first music society, set up in 1772 with three hundred members, drew large crowds to its weekly concerts during its five-year existence. Each member paid annual dues of ten rubles, part of which went to support the society's orchestra. A second music society founded a year later gave concerts every Saturday night and held monthly dances and masquerade balls during the winter months. The society's five hundred members each paid fifteen rubles to fund a fifty-member orchestra and occasional soloists, along with rent for the home of N. I. Chicherin, St. Petersburg's head of police, where they met. After the society's demise in 1792, three wealthy residents of the capital (Messrs. Demidov, Sikstel', and Bland) established the third music society with approximately four hundred members. Although this organization lasted only four years, not long after its collapse the beginnings of the Philharmonic Society were laid in 1802.[116]

Of all the city's social clubs, best remembered is the St. Petersburg English Assembly, or Club, which developed out of the informal gatherings of a group of foreigners at one of the city's hotels. When in 1770 the Dutch hotel owner was forced to sell his establishment, these European expatriates had to find a new home for their gatherings. Among their group was Francis Gardner, an English merchant, who suggested to his comrades that they use the opportunity to form their own "special, independent society, or club." After a unanimous vote, the St. Petersburg English Club, so named in honor of the majority of its members, was established on 1 March 1770 with fifty original members under the motto "Concordia et Laetitia."[117] The English Club eventually became one of the most prestigious and renowned clubs in all of Russia, thriving for over a century and including among its members the historian and writer N. M. Karamzin, the poets A. S. Pushkin and V. A. Zhukovskii, and the statesman M. M. Speranskii.

The club grew rapidly following its inception, and in little over a year its membership had quintupled, remaining around 250 for the next several years. By 1780, so many candidates were seeking admission that the club decided to raise the membership limit to three hundred. The English Club became a favorite spot of the city's better sort during the final two decades of the century and well into the next, when membership in the club was an unmistakable sign of social status. A truly international body like so many of the city's other sodalities, the English Club was filled primarily with Englishmen, Germans, and Russians who, while not among the founding members, quickly began to fill the assembly's ranks. Some of the most notable Russians active in the early years included the dramatist V. I. Lukin, the historian I. N. Boltin, the writer A. N. Radishchev, and the noted diarist and private secretary to Catherine II, A. V. Khrapovitskii, as well as representatives of prominent noble families such as the Golitsyns, Iusupovs, Odoevskiis, and Volkonskiis. But most of the Russians, like the members of other nationalities, were of less lofty status. Membership in the English Assembly, like that in most other societies and clubs at the time, was a solely male prerogative, denied to all women.[118]

After several initial moves in search of the best accommodations, the assembly finally settled in 1778 near St. Petersburg's Red Bridge in the residence of Countess Skavronskaia, which the members rented for thirteen hundred rubles a year. Here the English Club maintained a congenial atmosphere where its members gathered informally for a range of social activities "at any time of day or evening." Members were kindly requested to leave before midnight, however, to allow the head butler and servants a chance to clean and rest. In the reading room, the men could quietly peruse the numerous local and foreign newspapers to which the assembly subscribed; those in search of more lively pursuits could frequent the billiard room or join in a game of cards, although all "games of chance" were strictly forbidden. A wide selection of beverages was kept on hand in the buffet, and dinner was served by the club's own kitchen staff several times a week. Along with these daily affairs, the assembly also put on great feasts and dance balls on certain evenings.[119] As a result, the cost of running the English Club was substantial: by 1800 the annual expenditure had risen to forty thousand rubles, a fivefold increase from 1770. Completely self-financed, the club was funded largely by the members' yearly dues, which rose from fifteen rubles in 1771 to fifty rubles in 1795. Not all of the dues went to support the assembly, however, and beginning in 1772 annual gifts to the needy were made from a "special cash-box for the poor."[120]

Given the impressive size of its membership and the considerable amount of money involved in the club's operations, it is not surprising that the assembly, like similar societies of the age, drew up its own statutes to regulate its affairs as well as its members' actions. According to these statutes, consisting of twelve short articles, the club was to be administered by six officers whom the members would elect for six-month terms. The article on the induction of new members required that every candidate first find a sponsor among the membership who would then present his candidacy to one of the officers. The prospective member's name along with that of his sponsor would be displayed on a special board in the club for one week, during which time other members had a chance to inform themselves as to his character and personal qualities. The statutes stipulated a quorum of thirty members for each ballot of candidates—held at 8:00 P.M. on the first Thursday of every month—at which time a simple majority was required for admission. Besides members, only certain foreigners visiting St. Petersburg were allowed access to the club, and to make sure that all others were kept out, members were given a special sign or ticket by the butler that had to be produced upon entrance.[121]

The statutes of the English Club also evinced a concern for minimizing conflict and an interest in inculcating responsible and "proper" modes of behavior among the members. Those caught removing newspapers from the club, for example, were fined five rubles for each paper. In the case of disagreements or serious quarrels, each member involved was to select one or two mediators who, after having looked into the matter, would propose a settlement to which both parties were required to submit. If in the highly regrettable situation that a member "through his behavior offends the entire society," then he, "as a transgressor against peace and order," must be expelled from the club. All of these affairs and decisions were to be duly recorded in its regular minutes.[122]

Even if club life did not attain the same vibrancy in Russia's other capital, in Moscow too the final decades of the century witnessed the birth of several analogous sodalities. In 1790, following the great success of St. Petersburg's English Club, a group of Russian noblemen and merchants along with several foreigners set up their own English Club on Znamenka Street before moving to the home of Prince Gagarin a few years later. Like its more famous counterpart, the Moscow English Club offered members a pleasant setting to take their meals, to play a friendly game of cards or billiards, to read, or simply to chat. Furthermore, the club possessed its own formalized structure complete with bylaws and regulations, elected officials, and so on. The Moscow English Club did not share the longevity of its namesake, however, and was closed down in the reign of Paul I.[123]

A more successful undertaking was the Moscow Noble Assembly founded in 1783 by M. F. Soimonov and Prince A. B. Dolgorukii not long after the demise of a similarly named but short-lived Noble Club established three years before. After initially meeting in large private homes, the Noble Assembly eventually acquired its own building with its well-known columned hall where thousands of men and women from great aristocrats to petty nobles gathered every Tuesday night for balls, concerts, and masquerades.[124] Moscow appears also to have been home to several merchant clubs and a Foreigners' Club, where a German violinist and his wife gave a concert on 29 March 1780.[125] Finally, not to be outdone by St. Petersburg's Philadelphia Society, Moscow too supposedly had its own organized circle of debauchees, the so-called Eve Club, which met for two years in a foreigner's home in the German suburb's Poslannikovyi Lane in the early 1790s. At their weekly meetings men and women, including those from the highest levels of society, are reported to have engaged in "saturnalia of unprecedented vice."[126]

Clubs arose in Russia's provincial towns as well. In Kronstadt, Admiral S. K. Greig established among his officers the Maritime Society, part of whose function, along with offering a pleasant setting for relaxed sociability, was to further the "softening and amelioration of their [i.e., members'] manners."[127] Another new provincial society was Reval's Harmony Club founded in September 1792. The Harmony Club consisted of men from various social stations—all of whom possessed, their charter announced, exemplary "moral qualities"—and provided them an outlet for perusing the latest newspapers, journals, and books in its reading room, playing billiards and cards, and finally, attending the usual round of concerts, balls, and masquerades. The club's operations, and particularly the behavior of its members, were regulated through an elaborate set of bylaws that governed everything from electing members and officers to establishing the appropriate footwear to be worn during dances. Those guilty of violating any of the fifty-three separate rules were required to pay a fine (between twenty-five kopecks and twenty-five rubles), whose proceeds would be used "in aid of the poor."[128]

Along with these social clubs arose approximately a dozen learned societies and student circles devoted to improving society and to serving the homeland through the cultivation and propagation of various branches of knowledge and learning. The first and perhaps most influential of these institutions was the Free Economic Society, one of the oldest economic societies in Europe, founded in 1765 by several prominent courtiers and scholars as a new patriotic organization established to encourage the country's economy and agriculture. The Free Economic Society sought to achieve its goal of "promoting the happiness of the people" through the commissioning and publication of special reports and studies as well as its own journal, the *Transactions of the Free Economic Society*, that would place its ideas before "the universal scrutiny of the Public." In addition, the society put on public exhibitions and lectures in its library.[129]

Shortly after the founding of the Free Economic Society, several other learned societies dedicated to spreading enlightenment through the printed word were established in the capital. In 1768, Catherine organized the Society for the Translation of Foreign Books headed by V. G. Orlov, who was also the director of the Academy of Sciences, and administered by G. V.

Kozitskii, the empress's secretary. Attracting the participation of writers such as Bogdanovich and Ia. B. Kniazhnin, the society produced over one hundred translations of works ranging from the ancient Greeks to the French philosophes during its fifteen-year existence. In 1773, following the example set by the empress, Novikov formed the Society for the Printing of Books, which closed down the next year. Other learned assemblies included the Society of Lovers of Science, which published the *St. Petersburg Herald* from 1778 to 1781, and the National Free Society for Philanthropy founded in August 1785 by the minor poet and publicist F. V. Krechetov. Comprising some forty Russian and foreign men and women and dedicated to the advancement of jurisprudence in Russia, Krechetov's society appears to have met regularly until approximately 1793, when he was arrested during the period of fear and suspicion following the outbreak of the French Revolution. The capital was also home to the Society of Friends of Literature, a literary organization created in 1784 by former Moscow University students serving in St. Petersburg, and young officials, most notably A. N. Radishchev. It issued the journal the *Citizen Conversing*.[130]

The Society of Friends of Literature was directed by M. I. Antonovskii, previously a student at Moscow University where he had headed the Society of University Pupils. Founded in March 1781, the Society of University Pupils, continuing the tradition of student circles begun in the reign of Elizabeth with groups like the Society of Lovers of Russian Literature, had been created to enlighten students' minds and to reform their "depraved inclinations" through various literary projects and morally edifying speeches intended to make the students more useful to themselves and to the "beloved fatherland." One young member described the society's goals as to "succeed in the sciences, but to live according to the rules of good behavior."[131]

The Society of University Pupils had been initiated by the influential university professor J. G. Schwarz, whose other projects included the Pedagogical Seminar and, most importantly, the Friendly Learned Society. Like the Society of University Pupils, the Friendly Learned Society was intended both to raise the overall level of enlightenment in Russian society and, as a consequence of their labors to this end, to improve the moral character of its members. Thus, through its various activities—ranging from the publication of useful books, to the encouragement of sciences that were underdeveloped in Russia, to the financial and institutional support of Russian students—the Friendly Learned Society aimed to turn its participants into "the most useful and active" members of society as they promoted "the common good."[132] Although active informally since 1779, the society held its ceremonial public opening in November 1782 at the home of its wealthy patron, P. A. Tatishchev. The ceremony, which had been announced in the local press and with the publication of a special booklet, was a significant event in the cultural life of the city, attracting a sizable gathering including Z. G.

Chernyshev, the governor-general of Moscow and an influential Mason.[133]

But the Friendly Learned Society's initial success (its ability to attract powerful patrons and supporters and its high degree of public recognition), coupled with Schwarz's increasing reputation among educated society as the most dynamic of the lecturers at the university, helped to foster a feeling of resentment among many in the academic community. Perhaps most importantly, the university rector I. I. Melissino saw in the success of Schwarz's society a direct threat to the Free Russian Assembly, the group of professors, teachers, and students that the rector himself directed and had founded in 1771 as Moscow University's first learned society. Modeled after the French Academy, the Free Russian Assembly's main goal was to cultivate the Russian literary language; in addition, the society compiled a large number of documents of historical significance, which it published in the *Essays of the Transactions of the Free Russian Assembly* between 1773 and 1783.[134] The recollections of a former member, S. A. Tuchkov, describing how he joined the assembly suggest the social and psychological significance of membership in these bodies: "Such an offer was of course flattering for a proud young man beginning to enter society."[135]

Concerned that Schwarz's new organization was eclipsing his own, Melissino proposed that they merge their two societies. Schwarz rejected the offer and incurred the disfavor of the university's directorship, having already aroused the envy of a large portion of the professoriate. By the end of 1782, he was forced to resign from his position at Moscow University.[136] Still, the Friendly Learned Society, essentially reorganized as the Typographical Company following Schwarz's death in 1784, went on to enjoy a productive existence until 1791. Melissino's Free Russian Assembly, on the other hand, ceased operations prior to 1784. An attempt in 1789 by Melissino's successor I. I. Shuvalov to reinvigorate the assembly's activities through a new organization, the Assembly of Lovers of Russian Learning, proved unsuccessful.[137]

Even though such literary circles and learned societies were found almost exclusively in the capitals, a few appeared in Russia's provinces as well. Around 1788, for instance, E. A. Bolkhovitinov, a seminary teacher who went on to acquire some note as a theologian, organized a small literary circle in Voronezh, comprising academic colleagues, seminarians, state officials, and local gentry. All of the group's members had their own sobriquet, usually taken from the ranks of past Italian sovereigns, and they frequently gathered for evenings of conviviality as well as for more serious discussions of books on philosophy and science. In 1798, Bolkhovitinov's circle also played a decisive role in setting up a press in Voronezh, which produced a steady stream of works on a variety of subjects for the next two years.[138] And in 1759 in the northern port of Arkhangel'sk, A. I. Fomin, a local historian, ethnographer, and book merchant who later became a member of the Free Economic Society, founded, along with his colleague V. V. Krestinin, Russia's first historical society. During its nine-year existence, Fomin's society gathered numerous documents of historical value, many of which the society passed on to the botanist and academician I. I. Lepekhin during his visit to Arkhangel'sk in 1771.[139]

Finally, one more group of societies that might be described as professional organizations—although a lively sociability often figured prominently here too—came to life around this time as well. On 24 November 1784, for example, merchants in St. Petersburg founded the Mercantile Society as a place where they could gather both to discuss news related to their business affairs and to relax in a convivial setting. Occupying its own house on the fashionable English Embankment, this association went on to become one of the most respectable clubs in Russia, after the English Assembly.[140] In addition, St. Petersburg hosted two informal merchant societies— one English, the other German—about which little is known except that both put on balls every winter.[141]

In June 1775, fourteen residents of St. Petersburg led by Joachim Grot, the pastor of St. Catherine's church on Vasil'evskii Island, gathered to found the Funeral Society, perhaps best described as Russia's first life insurance company.[142] Open to men and women between the ages of twenty and forty and admitting *"people of all social stations,* except those serving in combat," the society grew rapidly. Within seven months its membership increased to almost 100, and by 1780 it counted over 250 past and present members.[143] While attracting some representatives of Russia's traditional elite, such as Count R. I. Vorontsov or the archpriest Vasilii Muzovskii, the Funeral Society drew the bulk of its membership from the merchant class, and then from the growing ranks of state administrators and officials. Among the occupations represented were silversmith, barber, court musician, broker, academician, professor, actor, and bookseller. Some two dozen married couples belonged to the society, including the leading court actor I. A. Dmitrevskii and his wife, A. M. Dmitrevskaia (née Musina-Pushkina), one of Russia's first actresses, along with many single women such as the widow of the court dentist. A truly international organization, approximately two-thirds of the members were foreigners.[144]

Admission to the Funeral Society cost ten rubles. On the death of a member, everyone was required to contribute two rubles to a general fund that would be presented to the deceased's heirs.[145] The entire membership elected a group of officers to administer the society; they in turn selected an aide and several messengers, whose "sobriety, faithfulness, [and] integrity" had to be beyond reproach, to help run the day-to-day operations. A system of monetary fines was set up to ensure that all members and officials complied with the society's rules and regulations. The collected fines were to be distributed to the poor.[146] To prove that its aims were free of any avaricious intent and that it was not some sort of elaborate scam, the Funeral Society disclosed all its actions to the public. In addition to the publication of its statutes, specific goals, and membership lists, the society also announced in advance all its general meetings and then published their results in the *St. Petersburg News*.[147]

. . .

For centuries European travelers have commented on what they perceived to be the vast social, cultural, and political gulf separating Russia from their own native lands. In *La Russie en 1839*, one of the best known of such accounts, the marquis de Custine described a country frozen in the icy embrace of autocratic oppression. As evidence of this oppression, he pointed to St. Petersburg's total lack of such amenities as Europe's prevalent coffeehouses—where the literate classes gathered to discuss the day's affairs—and even the absence of regularly printed gazettes and journals on which such discussions were based. Custine drew a picture of a cowering and seemingly formless society, one devoid of clubs and assemblies, newspapers and other periodicals—of the very institutions that characterized most European societies.[148] It is difficult to exaggerate the impact of Custine's work on Western perceptions of Russia during the past 150 years. So powerful a spell has Custine cast over Westerners that even as late as the 1950s, members of the American Embassy staff in Moscow read him as a guide to unlocking the mysteries of the inscrutable and (reportedly) immutable Russian soul. Undoubtedly, Custine's brilliant work contains some undeniable truths and unforgettable aperçus, as any reader who has ever visited Russia knows; yet even to such a favorable critic as Irena Grudzinska Gross, *La Russie en 1839* is a deeply flawed book, burdened by clichés, imagined conversations, and the author's pronounced Russophobia. Indeed, Custine's Russia, she remarks, springs less from a desire to understand this land than from the anxieties produced by the unspeakable violence of the French Revolution. In traveling to Russia, Custine sought to assuage these anxieties, to reassert his own identity as a member of a purportedly civilized Europe by safely locating barbarism beyond its borders among the despotic steppes of an imagined Russia.[149]

Ultimately, whatever the accuracy of his portrayal for the Russia of Tsar Nicholas I, Custine's description is clearly at odds with the picture of Russia in the previous century, and particularly with the era of Catherine the Great. During that age the women and men of Russia's educated classes saw themselves as belonging to a public, a distinct social formation within Russian society, bound together by the burgeoning print market and the increasingly complex network of salons, theaters, clubs, and societies. The recognition of a public sphere, or civil society, in eighteenth-century Russia points to the overlooked similarities between Russia and the rest of Europe and raises the question of how this social formation compared to those emerging at that time in other states. Unfortunately, the comparative history of civil society in early modern Europe has yet to be written, and definitive answers to this question cannot be given. The situation is compounded by the fact that what has been written has focused more on the idea of civil society than on its actual, concrete manifestations.[150]

Nevertheless, a few general features can be noted that suggest the ways in which Russia's civil society differed from those found in other parts of Europe. Perhaps most apparent are the greater numbers that made up the public and the greater density of the institutional networks that formed civil society in the West. One sees this in the fact that in the eighteenth century as many as twenty thousand Londoners gathered nightly in local clubs, or that France was home to some one thousand Masonic lodges, or that approximately fifty academies and literary and learned societies met on German soil.[151] The relatively small size of Russia's civil society in the eighteenth century need not be viewed as a weakness, however, since for classical theorists of civil society, men like Adam Ferguson and Montesquieu, the concept referred to a small society held together by the mutual trust possible only through personal acquaintance.[152] Such a state of affairs was actually more realizable in a setting like Russia where civil society exhibited a more personal, familiar character than it did in other European states.

The greater informality and intimacy of Russian civil society suggest another difference. Civil society requires the existence of mutually constitutive public and private spheres, and it is precisely in the eighteenth century, and most noticeably during the reigns of Peter I at the beginning of the century and Catherine II at the end, that this distinction began to take definite shape in Russia. Yet, the bifurcation of public and private did not progress as far there as it did in places like England or France, and throughout the century Russian state and society remained more interwoven and entwined than was the case in much of Europe. One sees this in the degree to which the exercise of power at all political levels, as has frequently been remarked, was resistant to attempts at bureaucratic formalization and regulation and continued to present a personal face, what John LeDonne has referred to as "the privatization of public power." One also sees this in the undiminished power of certain private individuals to dominate the public sphere. The formidable competition that the Sheremetevs' personal theaters in Moscow and Kuskovo created for public theaters run by men like Maddox reflected this condition as did the fact that much of public life in Russia remained centered around the private palaces and homes of the country's leading families, a tradition that lived on into the next century.[153]

According to Elise Wirtschafter, the lack of an unambiguous demarcation between public and private, state and society, was reflective of a more general "absence of clear demarcations" within Russian society, of its relative undifferentiation and borderlessness, both in terms of its physical geography, where town and country appeared to merge imperceptibly together, and its social groups, which were only marginally distinguishable from each other.[154] Still, it would be wise not to posit too stark a contrast between state and society for the rest of Europe either. As Adam Seligman correctly noted, in the eighteenth century, civil society was originally envisioned as "a realm of social mutuality," and it was only at the very end of the eighteenth and especially in the nineteenth century that it came to be chiefly understood as a realm beyond the state.[155] In the eighteenth century, this realm of mutuality was the home of the men and women who constituted the elite, and it was defined more by its openness to those individuals with the prerequisite manners, morals, and means, and less by its separateness from the state. This aspect of the idea of civil society has generally been overlooked by past observers who, by adopting a vague Tocquevillian notion of civil society linked to a rising middle class divorced from and largely distrustful of the state, have detected the development of a Russian civil society only in the final decades of the nineteenth century.[156]

Also crucial to understanding Russia's civil society is the question of gender and the place of women in this sphere. While historians of western Europe debate whether the Enlightenment and the new public sphere were liberating moments that created novel opportunities for women, or whether these developments gendered public space in favor of men and restricted women to a private zone of female domesticity, there can be little doubt that the birth of the public sphere in Russia signaled the inclusion of women in society.[157] The rise of society in Russia created for a small though growing number of women a new arena for cultural, intellectual, and social activity as they took their place alongside men in the country's salons, ballrooms, theaters, pleasure gardens, and societies. This was the case not only for those truly exceptional women such as the empresses Elizabeth and Catherine II or Princess Dashkova, but for more modest figures as well. The poet E. A. Kniazhnina, for instance, while still a teenager published in the *Busy Bee* (the journal of her father, A. P. Sumarokov) and hosted a lively salon for St. Petersburg's literary and artistic circles with her husband, the writer Ia. B. Kniazhnin. The writer, poet, and translator M. V. Khrapovitskaia-Sushkova was a member of the Kheraskovs' salon and contributed to several journals. The actress A. M. Dmitrevskaia belonged to St. Petersburg's Funeral Society.[158] While these women cannot be viewed as typical of the age, their lives suggest the new range of social and intellectual possibilities for public action then emerging in Russia. Of course, this is not to say that patriarchy was overcome, or that Russian women achieved legal equality. Nor does it mean that women were accepted everywhere within Russian civil society, as evidenced by the practices of numerous societies, including the Masonic lodges. Indeed, the changing relationship between the sexes may well have unleashed a backlash among those who felt that things had gone too far. Playwright D. I. Fonvizin's popular comedy *The Minor* (1782), for example, has been interpreted as an attack on women's enhanced status and a defense of traditional Russian patriarchy. But the public sphere allowed a variety of voices and opinions, and along with *The Minor* educated Russians could also read articles like those by P. S. Svistunov in *Idle Time for Good Use* (1759–1760) that argued in favor of female education and its benefits to society as a whole.[159] The development of a public sphere meant that questions about the rightful place of women in society were being openly discussed and debated, and the conversation was not solely among men.

Although civil society in eighteenth-century Russia remained a fragile construct, its existence cannot be denied and must be borne in mind when considering the specific social setting that was home to Russia's Masonic lodges and the world to which the Masons belonged. Many of the same men who frequented salons and theaters, who joined literary societies and social clubs, who subscribed to journals and patronized the cities' booksellers also

M. M. Kheraskov (1733–1807) was a rector of Moscow University and a prominent poet and writer. He belonged to several lodges and to the Rosicrucians and penned two Masonic allegories—*Vladimir Reborn* (1785) and *Cadmus and Harmony* (1789).

belonged to the Masonic movement. Salon hosts such as Kheraskov, Shuvalov, and Shcherbatov were active and prominent Freemasons. S. K. Greig, the founder of Kronstadt's Maritime Society, belonged to that city's Neptune lodge for several years, quickly advancing to the Master degree and even serving as its Grand Master for a time in the 1780s. Before establishing his Philadelphia Society, P. I. Melissino founded the St. Petersburg lodge Silence, where he acted as Grand Master throughout most of the 1770s, and to which he still belonged as late as 1787. His brother, I. I. Melissino, founder of the Free Russian Assembly, was a member of the same lodge from 1786 to 1787. Schwarz, Novikov, and P. A. Tatishchev, important figures in the Friendly Learned Society, all played large roles in Russia's Masonic lodges as did less well-remembered Freemasons such as A. I. Fomin, the cofounder of Russia's first historical society in Arkhangel'sk, who served as an officer in that city's Holy Catherine lodge and in 1787 helped to establish there the North Star lodge. Nowhere was this state of affairs more visible than in the St. Petersburg English Club, the vast majority of whose members during the first decade of its operation belonged to local lodges.[160]

Yet while the Masons, both as individuals and as a group, formed part of the public, sharing its various social activities, cultural practices, and overall mentality, they did not view their membership in the order in the same way as they did these other aspects of their lives. Attending a ball, enjoying the diversions of the vauxhall, or even joining a literary society did not carry the same symbolic weight as belonging to the international brotherhood of the Freemasons did. Why? Part of the answer has been revealed by the examination of life in the lodges that showed the great personal as well as social value the brothers attached to "working the rough stone." Another part has to do with how the brothers conceived of themselves as Freemasons and of the lodge as a symbolic space. For, although being a Mason implied possessing the requisite social skills and intellectual habits to move with ease and confidence among the men and women of the public gathered in the country's drawing rooms, lecture halls, and pleasure gardens, at the same time it also implied possessing something more, something better, that distinguished the Masons and their assemblies from these other people and places.

3

Virtue's Refuge, Wisdom's Temple

. . . and he, having separated himself from the world's masses, can say with
a sense of his own joyous distinction: I am satisfied.

—*From My Work on the Rough Stone*

CONSTRUCTING MASONIC SPACE

"Freemasonry," according to one anonymous Mason,

> sees in all people brothers to whom it opens its temple in order that they may
> be freed from the prejudices of their homeland and from the religious errors of
> their forebears, spurring them to mutual love and aid. It does not hate or per-
> secute anyone, and its purpose can be defined in the following manner: to ef-
> face among all persons the prejudices of caste, of conventional distinctions of
> birth, of opinions, and of nationalities; to destroy fanaticism and superstition;
> . . . to further with all its powers the common good and in so doing to create
> out of the entire human race one family of brothers, bound by the ties of love,
> knowledge, and labor.[1]

With these words the unknown Freemason captured one of the central
tenets of the Masonic movement—love for one's fellow man, a sentiment
repeatedly emphasized in the order's bylaws and regulations. The Common Institutions of the Freemasons, for instance, listed as chief among its
obligations "true love toward all people," and the Exegesis of the Statutes
or Regulations of the Freemasons reminded each brother that "[e]very suffering and grieving being has a holy right over you" and even cited the
Golden Rule, instructing all Masons to "[l]ove thy neighbor as thy self."
Free from the narrow constraints that segregated humanity, the Mason, as a
true cosmopolitan, was at home everywhere in the world: "The universe is
the Freemason's homeland, and nothing characteristic of man can be
foreign to him."[2]

As students of the movement have long noted, Freemasonry's affirmations of love for all mankind and its call for a universal brotherhood of man marked the movement as truly new and modern when placed in the context of the ancien régime. Through its explicit disavowal of the traditional institutions of corporation, estate, and religion, the Masonic lodge constituted an important bridge to the modern world. Still, it would be inaccurate to focus attention only on protestations of unqualified love for humanity as the ultimate expression of Freemasonry. Indeed, upon further examination a much more complex and contradictory picture of the Masons' stance toward their fellow men becomes apparent, for entwined with a loudly proclaimed love for humanity was an equally strong sentiment of superiority— occasionally even contempt—toward the rest of the world. While the Freemasons may have seen the entire universe as their homeland and acknowledged a common divine spark animating all of humankind, this did not mean that everyone was welcome into the order, or that all men were essentially the same—indistinguishable and equal. The Masons' conscious rejection of certain existing social hierarchies did not necessarily entail a complete rejection of hierarchy per se, and present throughout Masonic discourse and practice was an undeniable inclination toward mapping and classifying the world according to their own standard. Freemasonry was less a repudiation of hierarchy as a principle of social organization than an attempt to fashion a new one.

On the most fundamental level, the Masonic urge to classify manifested itself in the division of the world into two unequal halves: Masonic space and non-Masonic space. The lodges represented but one strand in the larger fabric of civil society that took shape in the eighteenth century. Freemasonry, as both a physical institution and a discursive formation, helped to constitute and to define the public along with other newly established social organizations, ranging from salons and lecture halls to student circles and commercial societies. What distinguished the Masonic movement, however, was its self-described relationship to the rest of society. To the Masons, this larger sphere was wicked and vice-laden, whereas the lodges represented islands of light and virtue that had been carved out from within non-Masonic space. According to the Exegesis of the Statutes, nature's original vow of equality had long ago been destroyed by man with the advent of false and artificial divisions among humanity.[3] Thus, while the lodges were not originally intended to be "immured and unknown in society," they were sadly forced to adopt a less open position when "violence, guile, and malice prevailed" throughout the world.[4] In the Masons' view, society was populated by human types unlike those found in the lodge. Usually referred to as the "profane" or the "unenlightened," these types were held to be grossly inferior people who remained unsuspecting captives of their drives and passions—literally, personifications of the "rough stone." When viewed as a whole, the "natural, unrefined man" formed a "multitude . . . of the profane wandering in darkness."[5] In his encounters with the profane, the Mason was told to "set a good example for him through his actions and never to disdain him."[6] Superior mores, the Freemasons maintained, would make every brother "respectable in the eyes of the unenlightened."[7]

The lodge, on the other hand, existed as a unique space distinct from the rest of the larger social order. It was founded upon and operated according to norms and principles that stood in radical opposition to those outside the lodge. In direct contrast to the perceived vanity and vice of society, the lodge was constructed as a "refuge of Virtue" and a "school of truth." Infused by a spirit of "absolute harmony," it represented a "majestic edifice appointed to strengthen the highly weakened bonds of humanity and morality."[8] The lodge figured as a site of virtue and knowledge amid a chaotic void of vice and ignorance. Given the loftiness of Freemasonry's stated mission and the assumed depravity of society at large, Masonic space took on a marked and sacred quality setting it apart from the space occupied by other clubs, societies, and assemblies. The prohibitions in the Masons' General Laws against the establishment of any lodge "in an inn or any other public place" and against discussing any aspect of Freemasonry "in public gatherings, or on the street" attest to the special quality that necessitated a clear division between the world of the mundane, on the one hand, and that of Masonry, on the other.[9] Masonic statutes and regulations made a similar point, asserting that the lodge was not simply another public space but that it operated by its own rules distinct from those governing the larger social order. While the Masons did not reject society's norms outright, they simply did not consider them to have any place within the lodge. The Common Institutions of the Freemasons warned the brothers:

> But beware not to introduce to our temples obsequious marks of distinction that we do not recognize, leave your merits and signs of ambition outside our doors and come in to us with only one companion—your virtue. Whatever your worldly title, in our lodges yield to the most virtuous, the most enlightened.[10]

And in a similar vein the Statutes or Regulations of the Freemasons commanded each member to

> Respect society's lawful hierarchies: in our temples we know none except that which distinguishes virtue from vice. Beware not to introduce any changes capable of violating [this] equality, and when before outsiders do not be ashamed of the honorable man who whilst amongst us you had kissed as a brother.[11]

Even though the Masons professed to recognize only the markers of vice and virtue in judging a man's worth, as with other social sites in the public sphere, the lodge admitted to an equality only of the educated and relatively well-to-do, and its doors were kept closed to the "rabble." According to the *Free-Masonic Magazine*, a collection of Masonic speeches, discussions, and songs published in Moscow in 1784: "It is ignoble and unjust to think of Masonic lodges as one does of the feeble and unthinking mob . . . the Freemasons' lodges are barred to no one except the mob. While closing their doors to the weak, the evil, and the depraved, they open them without exception to all meritorious and distinguished men and particularly to the man of virtue."[12] Along the same lines, the bylaws stipulated that if individuals lacking personal freedom (for example, serfs) were admitted to the lodge, it be only as servant brothers.[13]

In addition to the rabble, the Masons also barred their doors to women. Freemasonry, like any fraternity, had originally been conceived as an organization for men and so it remained, despite some important exceptions. The chief justification for keeping women out of the lodges centered on the belief that the presence of females would distract the brothers from their important work and weaken their fraternal bond. Upon initiation new brothers were informed: "True, in our gatherings here we are denied their presence, so that their alluring beauty will not turn us into rivals and give birth to jealousy and discord."[14] On one level, this justification can be read as an admission of the brothers' weakness. To recognize women's power over them was to admit failure in curbing their base passions; still lacking the strength of character and degree of self-control necessary to resist temptation, the Masons were forced to deny women access to their meetings. Russian Freemasons sang amongst themselves: "Admit and be agreed with us, / You delicate love, / That you control our hearts / And inflame our blood."[15]

Of course, to recognize the power of women to control the Masons' hearts and to inflame their blood was at the same time to reiterate traditional male fears about women's ability to lead men astray. In the same song the brothers asserted that had Samson been a Freemason he would never have been deceived by Delilah. Thus, women had to be kept out of the lodges not only since their innocent charms could distract the brothers from their business, but, more cynically, since women's designs were possibly at odds with those of the brothers. On still another level, one detects in the Masons' writings a sense that women should be excluded from the order not because of their irresistible charms or because they might be out to steal the Masons' secret, but because they are simply incapable of behaving in a manner expected of Freemasons. In 1781, in his *Moscow Monthly Edition*, Novikov published an anonymous speech most likely originally delivered in an English Masonic lodge that captured this notion: "At the conclusion of exercises, we join together for the outpouring of our hearts' gaiety without the slightest pretense. Here among us there are no *chatterboxes*, no *slanderers*, or backbiters; our work is devoted to *silence*. And therefore we can say in a figurative sense it is carried out in secret places in which no cock crows, the voice of discord does not gain access, and the din, cry, and ungovernable wrath of women is never heard."[16] The misogyny here is unmistakable: women are termagants incapable of working the rough stone.

It is possible that the inclusion of women in Russian society beginning in the reign of Peter the Great produced a backlash among men who, whether consciously or not, wished to keep certain areas of social life for themselves, and that the Masonic lodges, which arrived in Russia with an existing ban against women, provided these Russian men with just such an all-male institution. Yet, if this was the case, Russia's Freemasons did an excellent job of camouflaging such designs, for their bylaws and charters, regulations and rituals, letters and speeches betray no preoccupation with the relationship between the sexes. In marking Masonic space and its inhabitants, the Masons emphasized their symbolic distance not from women, a secondary concern, but from other men. In the Masonic worldview, the gulf separating men from women, significant though it was, could, under certain circumstances, be bridged. The gulf separating the man of virtue from the profane could not. To quote the lines of a song sung by the men and women Masons of one adoption lodge: "Friendship among the common people is nothing more than a shadow, a disguise. Among us it is a sentiment, as solid as it is sincere."[17]

In relation to existing religious and state institutions, Masonic discourse placed the lodge in a superior, though not necessarily antagonistic, position, claiming to provide its members with a body of knowledge and a set of practices necessary for a kind of self-improvement the other institutions simply could not offer. Before new members could be initiated, their interrogators asked whether they had already sought virtue through "religion or civil law."[18] The fact that an individual sought admission to the lodges was understood by all Masons as a sign of his inability to locate virtue in these two other spheres. The implicit assumption shared by the Masons, therefore, was that the lodge was not simply *a* seat of virtue and wisdom, but *the* seat. As the inhabitants of this sacred space, the Masons viewed themselves as constituting a "universal fellowship of virtuous souls" and "the most highly esteemed brotherhood."[19]

Each Mason signaled his membership in the brotherhood by using the order's own "Masonic language."[20] Made up of unique words and phrases (for example, "working the rough stone") as well as physical and visual signs (such as the Mason's grip and numerous gestures along with the entire system of Masonic hieroglyphs), this language strengthened the brothers' sense of common identity by providing them with their own argot distinct from the general language of society. Moreover, Masonic language helped to highlight the border separating the Freemasons from the unenlightened. Outside the lodges, brothers could communicate their membership in the order through carefully chosen words or gestures; that such signs might be witnessed by the profane was of little concern—in fact, implicit in the possession of a Masonic language was the desire to make its existence known to the untutored. The use of unusual signs before outsiders conveyed the presence of a mysterious community to which these others did not belong and whose ways they did not understand.[21]

What sort of men were considered worthy of joining the Masonic community? The primary concern in admitting new members was that no one be accepted who might bring shame to the lodge or sully the name of Freemasonry. In one set of bylaws, the section on introducing potential brothers highlighted the importance of not recommending someone who might bring "dishonor" to the brotherhood: "A seeker, against whom a charge has been made public, or brought to court, that he holds dissenting ideas on Christian doctrine, or about whom it is rumored that he has committed flagrant and disgraceful offenses, or [possesses] human vices, shall be expelled from the Order by a single black ball." The chief requirements

A Masonic snuffbox. The significance Masons attached to membership in the order is conveyed by the large number of snuffboxes, watches, and other trinkets they had made covered with Masonic hieroglyphs. Carrying these objects around with them they could communicate in a discreet way their status as Freemasons to the rest of society. The cross-and-pelican (a symbol of self-sacrificing benevolence) suggests that the original owner may have belonged to the degree of the Knight of the Eagle and Pelican or was possibly a member of St. Petersburg's Pelican of Charity lodge.

for admission were proof "of a good . . . life" and of one's personal good "qualities."[22] When, in August 1783 at a meeting of the Golden Key lodge in Perm', Ivan Cherkasov proposed for membership Ivan Ashitkov, an official in the local administration, the brothers expressed unanimous support upon assuring themselves of the applicant's "good manners and morals and virtuous behavior."[23] These factors were deemed so important that the order's age requirements (twenty-one years for the sons of Masons and twenty-four for all others) could be waived for prospective members who exhibited "steadfast and noble comportment."[24] Even the Swedish Masonic system, introduced to Russia in the 1770s, that emphasized noble lineage as a prerequisite for admission recognized the greater importance of personal conduct and reputation. In a letter to his superiors in Stockholm in 1777, Prince A. B. Kurakin recommended an English merchant for their chapter since "[a]lthough Mr. Yaeger is not a nobleman . . . his probity is recognized and . . . he can help us to win over the English lodge, which is one of the better ordered here, one of the wealthiest, and whose members are all estimable and who can bring honor to the order."[25]

A song popular with Russia's Masons listed in concise fashion the necessary qualities:

> He who knows integrity,
> Who always observes the truth,
>> Who loves to live moderately
>> A Freemason he may be.
> He who shows his friends but honesty,
> Who knows the laws of camaraderie,
>> Who's honored to live in harmony,
>> A Freemason he may be.
> . . . .
> He who helps the poor,
> Who protects them from calamity,
>> Who acts as virtue's protector,
>> A Freemason he may be.[26]

Given the movement's lofty self-image, initiation into the lodges was accorded great symbolic weight. To join the brotherhood meant leaving the disordered world of the masses and passing into the realm of the elect. Becoming a Freemason required undergoing a rite of passage that transported the individual from one world to another, an initiation that like other rites of passage intended to signify, in the words of Mircea Eliade, "a radical change in ontological and social status."[27] In his "Sketches from My Life," Baron Heinrich Jacob Schröder, a Mecklenburg landowner who entered Russian service in the early 1780s and participated in Moscow's Masonic circles, deemed only three dates from his entire life worthy of mention: his birth on 8 July 1757; his initiation into Rostock's Three Stars lodge on 17 February 1776; and his receiving of the Master degree on 8 April 1778—"An important year in my history."[28] The most revealing case of the powerful significance attached to joining the Freemasons is that of I. M. Strazhev. A serf belonging to A. I. Verevkin, Strazhev received his freedom after being admitted to his master's Hercules in the Cradle lodge in Mogilev, taking the name of Strazhev from the Masonic office of guard (Russian—*strazh*), which he held. To Strazhev, who went on to become a merchant in Vitebsk, becoming a Mason literally meant becoming a man.[29]

The Masonic initiation ritual vividly conveyed the nature of the transformation from profane to Freemason.[30] Every new brother began his initiation in the lodge's "dark" or "black little temple," a small, dimly lit antechamber intended to serve as a representation of the unenlightened world beyond the lodge. Not only did the future brother find himself isolated in this dark chamber, he was also blindfolded as further demonstration of "our natural blindness and ignorance." Unable to look upon the world around him and cut off from all external stimuli, he was instructed to "direct his gaze inward upon himself."[31] As another sign of his forsaking the sphere of the profane for the sacred, the candidate had to remove his outer dress and hand over all metal objects (such as money, jewelry, medals, or similar insignia), attributes of the profane world that had no place in the one he was about to enter.

During the initiation ceremony—a "journey" in Masonic language—the candidate was brought progressively closer to the lodge's sacred center, to the light that illuminated the Masons' common work and that penetrated to the heart of each brother. In order to highlight the great symbolic distance separating the lodge from the outside world, and consequently, the enlightened brothers from the unenlightened masses, the candidate could not pass from profane being to Freemason without traversing several intermediate stages. He began the initiation ceremony as a "candidate"; in degrees he rose to the level of "desirer" or "seeker," next to that of "sufferer," followed by "petitioner," and finally, "brother." Each stage corresponded to a distinct physical location and signaled the progress the initiate had made in his journey.

Upon entering the small black chamber, the "candidate" advanced to the status of "seeker"; after being led out of the chamber and across the lodge threshold he became a "sufferer." Once inside, the "sufferer" had to circle the lodge interior three times—again intended to symbolize the phased progression toward virtue—before approaching the altar, the lodge's symbolic center, and taking the Masonic oath. Yet even then he remained blindfolded and somewhat apart from the brothers. To become a full-fledged member he had to be allowed to see the light that illuminated their works, which they permitted him to do, but only gradually, by degrees. Having proved himself worthy of Freemasonry's light, the "petitioner" was first allowed to see only the "weak light": the brothers removed his blindfold and before him on the altar burned a single candle. His eyes were covered once again and after all the candles in the lodge had been lit, the blindfold was removed for the last time, and he was finally allowed to behold the "full light." Now a "brother," his transformation was complete. He had been led carefully from a state of blindness and ignorance to one of sight and self-knowledge. No longer part of the profane crowd, he now belonged to a select order, to a society "celebrated and renowned for its virtue, friendship, loyalty, and honesty."[32]

In the lodge the new brother found a spiritual home among men of like sentiments and character, bound together by a common pledge of loyalty and friendship. Just as his fellow Masons scattered throughout the world could rely on him in time of need, so too could he put his trust in the "sacred law" that obligated them to extend freely their help and friendship in return. As another link in this "chain of united hearts," the newly inducted brother was an equal partner in the Masonic community. The "sacred knot" that tied them together leveled all prejudices and distinctions constructed upon national origin, Christian denomination, and social standing.[33]

The Masons' attempt to claim the lodge as the lone site of virtue and enlightenment reflected a larger transformation in the country's moral geography over the course of the century. This transformation can be seen in the growing tendency to locate virtue in such intimate venues as the salon and the literary society, whose members—by emphasizing the social and personal utility of their endeavors—strove to distinguish themselves from the flatterers at court and others who frittered their lives away in endless games and idle chatter. It can also be seen in the Arcadia of the country estate that increasingly figured in the noble imagination as a refuge from the wicked ways of court and city,[34] or even, by the end of the century, in the image of a romanticized peasantry, noble, virtuous, and untainted by the corrupt civilization of its social betters.

Degrees and Systems

Although the Masons depicted the lodge as a realm founded on the equality of the virtuous, the drive for distinction that marked this space and its inhabitants as separate from the rest of society did not stop at the doors of the lodge. On the contrary. The same concern with classifying and segregating people according to the scales of virtue and knowledge was as present within the lodge itself as it was within the broader confines of the entire Masonic movement.

The most apparent distinction inside every lodge was the division of the membership into several degrees or grades: Apprentice, Fellow Craft, and

Assembly of Freemasons for the reception of apprentices. The candidate is led in blindfolded and free of all metal objects with his left breast and right knee exposed. At the opposite end, surrounded by the rest of the lodge brothers, sits the Grand Master, before him the carpet of the Apprentice degree.

Master. Initiates found themselves above the rest of humanity, but below many of their new lodge brothers in the Masonic hierarchy. Each degree reflected one's relative position on the path from vice and ignorance to virtue and knowledge. Expressed in the idiom of Masonry, each degree corresponded to a different state in the progressive working of one's inner self. At the lowest level was the "UNPOLISHED or ROUGH STONE," followed by the "STONE WORKED INTO A CUBE or READY TO BE USED IN THE EDIFICE," and finally the "DESIGN, that is, THE PLAN FOR THE EDIFICE."[35] The distinctions among the degrees were communicated and reinforced in various ways. First, every degree had its own set of Masonic attire and regalia that immediately identified a member's standing within the lodge hierarchy. Second, access to different lodge meetings was based on one's degree. Only Master degree Masons possessed the right to enter the gatherings of each of the three grades, a privilege that was communicated by the small ivory key worn around their necks.[36] Third, the different initiation prices expressed the inherent inequality among the degrees: according to the General Laws, initiation into the first two degrees cost half as much as initiation into the Master degree (fifteen as opposed to thirty rubles).[37] And finally, the various laws and statutes instructed those of lower degrees to honor and obey brothers of higher degrees for these brothers, having advanced further on the path to virtue and enlightenment, had much to teach their brothers.[38]

With greater authority and prestige, however, came increasing demands and expectations. Upon joining the Fellow Craft degree one had to pledge to "increase the zeal and attention" that had been shown as an Apprentice. To help him in this task, the initiate gave up the plain Apprentice trowel he had worn on his apron for a new "polished trowel" intended as a reminder that from then on his work "must be nobler than it had been as an Apprentice."[39] Along with a "gold trowel," every new Master degree Mason received a pair of spotless white gloves as signs of his "chaste and pure" conscience and as prompts to ensure that he did nothing to sully his conscience or make him unworthy of these additions to his Masonic wardrobe. As a Master he was expected to "distinguish himself from others through his virtue and knowledge."[40]

Along with the division into degrees, the lodge also possessed a hierarchy of Masonic offices. At the top stood the Grand Master. The fact that no direct communication was permitted between the Grand Master and the brothers suggests the extent of his authority and status. Rather, communication in both directions had to pass through either the Deputy Master (if the lodge had one) or the two Wardens, who acted as mediators. The Grand Master and the Senior and Junior Wardens composed the nucleus of the

Assembly of Freemasons for the reception of Masters. Laid atop a coffin on the Master carpet, a cloth the color of blood draped across his head, the candidate undergoes a symbolic death. The brothers point their swords at him in a threatening manner intended to ensure future loyalty. Three more Fellow Craft brothers await their turn. Whereas in the Apprentice initiation there were three candles, or lights, now there are nine in order to symbolize the Masters' greater enlightenment.

Assembly of Freemasons for the reception of Masters. The Grand Master raises the candidate out of the grave as he grants the Master Word, thus marking his rebirth as a Master Mason. The lodge officers can be identified by their jewels: in the center, the Grand Master wears a square; on the left, the Senior Warden wears a level; second from left, the Junior Warden, a plumb-rule; and next to him, the Secretary sports cross-quills on his coat. In addition, they all exhibit the obligatory Masonic apron.

lodge; these three constituted the "formation of a lodge." The addition of the Secretary and Orator marked the existence of a "just" lodge, and all seven officials formed a "perfect" lodge.[41] Just as each degree possessed its own regalia, so too each officer received his own special insignia as a sign of his unique position in the lodge.[42]

In addition to degrees and offices, several other classifications and definitions were used. Servant brothers, who found themselves at the very bottom of the lodge hierarchy, represented one such group. At the opposite end were the lodge's "honorary members," a special status offered only to a few distinguished personages. One more type of brother was the "visitor," that is, a Freemason visiting a lodge to which he did not belong. Although attending different lodges was permitted, and often encouraged,[43] the Masons evinced apprehension that only appropriate visitors be admitted. First, it was important to make certain that no profane individual claiming to be a Freemason gain entrance. Thus, all visitors had to be examined by the Master of Ceremonies and the Orator on their knowledge of Masonry before being admitted. If the examiners discovered that the visitor was indeed a "deceiver," then the Orator was instructed to tell him to leave immediately or else prepare himself for "unpleasant consequences."[44]

The fear was not simply that the uninitiated might furtively make their way into the gathering. In their wariness toward visitors, the Masons also singled out for exclusion every "spurious Freemason" and every "false brother." For just as the Masons separated themselves from the unenlightened by dividing the world into two unequal halves, they also drew a line down the middle of the movement itself between "true" and "false" followers and between "orthodox" and "unorthodox" lodges.[45] The opposition between "true" and "false" Masonry reflected the divided nature of a movement that since its inception had been a protean and heterogeneous phenomenon, both as an organization and as a set of ideas. Indeed, soon after its establishment, Masonry split and, throughout the 1700s, continued to divide into distinct rites and systems, each professing its self-appointed status as the only true form of Freemasonry. This was equally true in Russia where numerous Masonic systems vied for dominance during the eighteenth century. Although past scholars have thoroughly documented the complex and intricate history of Russia's Masonic systems,[46] they have not

A Masonic apron. The white leather apron was the first gift bestowed on the newly initiated brother, which he was required to wear at all their assemblies. Its color symbolized the purity of the order and was intended to be a constant reminder to him of the necessity to keep his manners and morals pure. Along with the usual Masonic signs, this apron has several symbols common to the Master degree such as the tears and the sprig of acacia used to mark the grave of Hiram Abiff, the architect and builder of Solomon's Temple.

examined the logic informing the desire to claim possession of "true" Masonry and how this desire fit into the larger project of Freemasonry.

When considering Masonry's heterogeneous nature it is important to note that the various Masonic systems were not as different from each other as has generally been thought. First, all Masonic systems in Russia—from the English variants of the two Elagin unions, to the alliance of lodges adhering to the Swedish system, and even to the Rosicrucians—had higher grades along with the initial three Masonic degrees. Second, all Masonic systems were arranged in hierarchical fashion, with one lodge or select group at the top of a particular alliance that claimed to possess ultimate authority. Third, access to these supreme bodies was limited to those of the highest degrees, a privilege that only members already in possession of such distinction could bestow. Regardless of the objective differences that existed between Masonic systems, the concern with distinguishing "true" from "false" Masonry is perhaps best seen as another manifestation of the desire for social distinction and exclusivity so prevalent in eighteenth-century Freemasonry. For, just as Freemasons were intent on erecting a barrier between themselves (as the guardians of virtue) and outsiders (whom they viewed as the personification of vice), so too were they committed to defending their purported monopoly on what constituted Masonry's essential truth. The degree to which one Masonic system diverged from another mattered little: the existence of other systems stood as an implicit challenge to the validity and authority of every branch of the movement.

The First Elagin Union, an alliance of Russian lodges headed by I. P. Elagin in the early 1770s and operating according to an English system of Masonry, exemplified such a practice. While members of this union could visit all affiliated lodges, they were prohibited from interacting with any that were outside their alliance. Brothers of the Urania lodge (part of the First Elagin Union) swore in the initiation oath their love and complete devotion to the Grand Master and promised never to join or even to visit any lodge that did not belong to their particular Masonic union.[47] This prohibition even extended to lodges and systems whose practices were largely in accord with those of the Elagin Union.[48] Primary importance was given not to defending or advancing a particular Masonic strain, but to ensuring discipline and unity within a particular alliance. Thus, at a meeting of Urania on 19 February 1774, V. A. Pashkevich was forced to apologize to the brothers for attending a gathering of the newly founded Equality lodge. Most revealing in this episode was the fact that the Equality lodge did not belong to a different system and had even requested (although not yet received) the official Masonic constitution from the Grand English lodge that headed the entire First Elagin Union.[49] Regardless of its teachings, since Equality did not officially belong to the Elagin Union it was perceived as a threat and was avoided by all loyal brothers.

Not all systems were perceived as equally threatening. The Elagin Union directed its most stringent prohibitions against the Masonic system led by Baron P. B. Reichel, founded in league with the Berlin lodge of Count Zinnendorf in 1771.[50] Initially, Reichel, whose lodges adhered to the so-called Weak Observance system, sought to join Elagin's alliance but was rebuffed. By the mid-1700s, after a short period in which the Reichel Union found it difficult to attract members, the alliance began to grow, gaining more and more Masons as rumors circulated that "true Freemasonry" resided there.[51] It was within this context of shifting Masonic allegiances that the lodges of the Elagin Union sought to impose greater control over their membership. The minutes of the Urania lodge from these unstable years display a pronounced zeal for limiting contact between the competing systems. Not only were those members of the Reichel system who wished to join Urania required first to submit in writing a statement renouncing their existing ties to Reichel's lodges, but even brothers who merely wished to visit the meetings of Urania had to make the same demonstration of loyalty to the Elagin Union. Yet once it became evident that the Elagin Union could no longer compete for members and entire lodges began transferring their allegiance to the Reichel system, Elagin and the remaining brothers had little choice but to admit defeat and join the ranks of Baron Reichel and his united lodges.[52]

The activities of the Second Elagin Union (formed in 1786) evinced the same concern with defending "true" Masonry. The statutes of this alliance required all brothers to inform their superiors of any unaffiliated lodge in which their "honorable works are profaned."[53] Such an occasion arose in 1787 when members of the Concord lodge began to operate in the eighth degree, a grade that until that time had been unknown in the lodges of the Elagin Union, which limited their works to seven degrees. The new practices shocked members of the Modesty and Urania lodges, who on 25 February 1787 drafted a lengthy letter to Elagin complaining of these new practices. In their letter, the brothers argued that these innovations were dangerous, that they did not form part of "regular and genuine Freemasonry" and that they "threaten[ed] the unity of the masonic order." The brothers maintained that contrary to the practice in their lodges, where higher degrees where extended to those who through hard work in the lower degrees had shown themselves worthy, this new development reflected the trend prevalent in other lodges where "they take advantage of young Masons' inexperience and for money, or out of reprehensible ambition, raise these novices into supposed higher degrees, thus swindling them and befouling the dignity of the order." What especially upset the brothers of Modesty and Urania was that the brothers who had been initiated into the new eighth grade adopted an air of self-importance and superiority, presenting themselves as privy to knowledge unattainable to the seventh-degree Masons who theretofore had occupied the top of the Masonic hierarchy.[54]

This last anecdote helps to clarify the development of ever higher Masonic degrees and ever more complex Masonic systems. The significance of this development for understanding the entire Masonic movement is revealed by a brief examination of three different groups within Freemasonry: Andrew's Masonry (or the Scottish degrees); the Theoretical Degree of the Solomonic Sciences *(Teoreticheskii Gradus Solomonskikh Nauk)*, also called the Theoretical Brothers; and the Rosicrucians. Whereas the conflict between the First Elagin Union and the Reichel alliance represented a struggle for predominance between two comparable Masonic systems, the tensions between the brothers of the Modesty and Urania lodges and the members of the Concord lodge reflected a dynamic in which a group of Masons did not claim to represent a separate Masonic system, but rather to have moved to a higher plane within a shared system, thus placing itself above the rest of the brotherhood. This latter type of relationship, one of superiority within a common Masonic movement and not total separateness from that movement, characterized Andrew's Masonry, the Theoretical Brothers, and the Rosicrucians in Russia.

Andrew's Masonry is the name usually given to several higher degrees within Scottish Masonry introduced in the 1730s in France and Germany. Created by the Scotsman Andrew Michael Ramsay (better known as the Chevalier de Ramsay), these degrees were later adopted by various Masonic rites including Baron von Hund's Strict Observance established in 1754, the Swedish Rite, and the Rite of Zinnendorf, the latter two founded in the 1760s by the German Johann Wilhelm Ellenberger (Count Zinnendorf).[55] Within these rites, several of which (for example, the Rite of Zinnendorf and the Swedish Rite) found broad acceptance among Russian Freemasons, the higher degrees of Andrew's Masonry followed directly after the lower three symbolic degrees (St. John's Masonry) common to all Masonic systems. These higher degrees, however, figured as something greater than merely the next level; rather, the Scottish degrees of Andrew's Masonry marked an important symbolic division within Freemasonry itself.

Initiates into the degree of Scottish Master, for example, learned that they had come to the end of "regular Freemasonry" and had reached a level that "not every Freemason can achieve," a distinction partially conveyed through separate meetings, modified rituals, and new regalia. The works of the lower degrees were described as essentially the same as those of the "unenlightened," the only distinction being that the Masons "of the three lower degrees devote and pledge themselves [to their works] in the Order's sacred temple, [while] the unenlightened are left to their own devices."[56] The new Scottish Masters, however, were soon to be entrusted with knowledge and secrets "of the loftiest and most vital" nature.[57] As a result, their duties and obligations would increase as well. It was no longer adequate that one simply exhibited "appropriate ethical or moral decency of bearing, words, and actions," traits expected not only of every Freemason regardless of degree but of every "honest man" in general. These qualities had to go deeper, they had to come from the very core of one's being: "Scottish Masters require not mere ordinary morals of outward movements, words, and deeds, but also sincere morals of unaffected purity, from which flow justice, truth, and all real blessings."[58]

A different set of Andrew's Masonry texts exhibits similar tendencies aimed at highlighting the superiority of the Scottish degree Masons. Unlike St. John's lodges, which generally sought to recruit as many members as possible, the maximum membership for a Scottish lodge was usually twenty-seven. No new members could be admitted except to replace vacancies brought about by the death or prolonged absence of a Scottish brother.[59] In the event of an opening, the Scottish brothers selected someone from among the Masons of a St. John's lodge "who distinguished himself through his abilities, fervor, and commendable deportment."[60] It was imperative, however, that the candidate be truly prepared for this important transition and that he had fully exhausted all the opportunities for learning found in the lower three degrees. For if the candidate had failed to attain this maximum level of self-knowledge and sought admission solely out of a "childish desire" to appease his sense of curiosity, his lack of achievement and dishonesty would quickly manifest themselves: lost in the rarefied atmosphere of the Scottish degrees, he would be unable to understand the teachings and fail to fulfill his new duties. Along with incurring "the ridicule of all learned men," the ultimate consequence of such a hasty and ill-advised action was the more horrific fate of eternal banishment from the order—the prescribed punishment for "such unfit members who simply could not improve themselves in the first three degrees" and for whom there no longer remained "any hope of correction." Accordingly, the candidate was repeatedly instructed during his initiation to examine "his very self in the most severe and impartial manner" since he was about to undergo a trial more demanding than any heretofore. Whereas in St. John's Masonry "they protected your weaknesses, here we expect you not to have any." With initiation into the Scottish degrees, the Mason was declared worthy of "greater enlightenment" and the "supreme knowledge" that came with admission into this "group of select brothers."[61]

Even though Scottish degree Masons enjoyed a status superior to the brothers active in St. John's Masonry, they could not lay claim to Freemasonry's highest ranks. In rites with Scottish degrees (Strict Observance, Swedish, and Zinnendorf among them), Andrew's Masonry merely represented one more level, albeit an important one, in a larger system of ever more intricate and finely graduated hierarchies, as the Theoretical Degree and the Rosicrucian order, introduced to Russia by way of Berlin in the early 1780s, make clear.[62] Whereas those in Andrew's Masonry claimed to admit to their ranks only the few Master Masons who demonstrated the utmost perfection in their works, the Theoretical Brothers, positioning themselves a rung higher, only accepted Scottish Masters, "who through their efforts have exhibited considerable fear of God, brotherly love, and desire for Wisdom."[63] For the Theoretical Brothers the primary barrier was not that separating St. John's Masonry from Andrew's Masonry; rather it was the barrier between themselves and all other Masons: "This Theoretical Degree . . . has been acknowledged and approved as a useful intermediary between Freemasonry and the higher internal brotherhood."[64]

As the number of degrees separating the lowest members of the Masonic order from those at its summit proliferated, brothers possessing higher degrees came to adopt an increasingly superior attitude vis-à-vis their lesser brothers. In a speech in 1789 to the Theoretical Brothers in Orel, Zakhar Karneev characterized lower-degree Masonry as little more than a "preparatory path" leading to "our high degree" and requested they pray that God give their lower-degree brothers the strength necessary to rise successfully through those levels so that they might someday join them in their privileged position.[65] As a prime example of the chasm of self-knowledge and wisdom separating them from those in the lower ranks, the Theoretical Brothers singled out the Masonic banquet, which they characterized as extravagant "bacchanalia" put on at considerable expense, with excessive drinking and abundant "victuals." The banquet (or, more accurately, the image of it the Theoretical Brothers sought to create) was offered as proof of the extent to which lower-degree Masons remained captive to baser physical drives and urges. The Theoretical Brothers, on the other hand, had all but eliminated banquets, restricting their occurrence to three Masonic holidays at which only bread and wine were served. Such demonstrations of restraint and self-discipline bordering on asceticism figured as signs of their greater virtue and enlightenment, of their progress in freeing themselves from the prison of their passions.[66] Their relatively small numbers—approximately sixty or so Theoretical Brothers in all of Russia as opposed to the twenty-seven brothers in each Scottish degree lodge—provided further indication of their unique status.[67]

Along with the Theoretical Brothers' increased sense of self-worth in relation to the rest of the brotherhood, there developed also among them a pronounced disdain toward the non-Masonic milieu. Whereas for St. John's Masonry, the uninitiated individual was simply the "profane" or "unenlightened" one, for those in the Theoretical Degree he became the "vain man": "His conduct is rapacious, wicked, malicious; his desires are lustful, impertinent, injurious; his thoughts base, gloomy, rude. In his affairs he is a brutal tiger or a voluptuous swine; in his cravings a lewd creature or insatiable beast."[68] As a whole the uninitiated formed a "common circle of slumberers" each of whom, living only for the external and fleeting rewards of power, wealth, and status, remained forever entrapped in the "gloomy pall" of the passions that kept him "unknown to his very self." The Theoretical Brothers, however, had freed themselves from this shroud, from the "din of passions" that held the multitude—as well as lower-degree Masons—captive.[69] Their advanced degree of enlightenment made them unique. They constituted a small group, according to Zakhar Karneev, that from among "millions like us" had been shown the "true door to wisdom."[70]

The door referred to by the Theoretical Brothers opened onto the order's final level, namely, the Rosicrucians, which had been introduced to Russia along with the Theoretical Degree. In fact, the two composed constituent parts in a unified Masonic system wherein the Theoretical Degree functioned as a link between Freemasonry's lower degrees and the order's "higher internal brotherhood."[71] The Rosicrucians formed the elite in a system that began at the lowly rank of Apprentice and ascended through the many degrees of St. John's Masonry, Andrew's Masonry, and the Theoretical Brothers all the way to this lofty terminus. Extremely selective, the ranks of the Rosicrucians were closed to all but a few. According to their bylaws, each Rosicrucian assembly, or "circle," was to consist of no more than nine members, a supposed "holy number" meant as a "sign of the end of all cultivated things," and a subtle allusion to their supreme status. Nevertheless, while the Rosicrucians constituted the elite within the order, there existed a hierarchy of several degrees amongst them as well.[72] Along with their exalted standing came heightened demands and expectations. Not only were the Rosicrucian brothers expected to carry out all the personal, social, religious, and state obligations required of every Freemason, but even the small amount of free time remaining was to be put to constructive use and not to be spent "in idleness or wasting time engaged in fashionable pursuits." Consequently, wherever the brothers found themselves—as long as they were not being observed by any of the profane—they were required to devote their conversations to those matters that further "[s]cience and wisdom, God's honor, and love for one's neighbor."[73]

The Rosicrucians' supreme standing in the order was reflected in their extreme condescension toward the milieu of the profane and lower-degree Freemasonry—an attitude seen earlier, albeit in milder forms, among Andrew's Masonry and members of the Theoretical Degree. While those active in St. John's Masonry had defined themselves as a community of the virtuous in opposition to a purportedly depraved society, the Rosicrucians projected a similar image of depravity onto the Masons themselves, placing them squarely within the profane sphere of the "brutish man."[74] Moreover, as further proof of their superiority within the Masonic movement, the Rosicrucians claimed the authority to confer at once all three St. John's degrees on any profane person they judged deserving.

Justification for such powers came in part from the fanciful history of the order put forth in the Rosicrucians' statutes that inverted the actual relationship between the Rosicrucians and the lower degrees. According to this source, the Rosicrucians did not develop from the Freemasons, but vice versa. They traced their origins back to the beginning of the world when the first "teachers of wisdom united and separated themselves from the unenlightened crowd."

> And in order that the superiors might better hide their true intentions . . . they established the three lower classes of so-called Freemasonry as a training ground for higher knowledge operating by certain parabolic adornments and rituals; and although throughout the remoteness of the ages they [that is, Freemasons of the three lower degrees] have committed many vain, deleterious, and strange acts, utterly profane and almost unrecognizable, still by brotherly right the most capable of subjects must always be elevated from this milieu, and no one, except Masters of the Light's Radiance, may join our circle.[75]

In "Some Facts on the F[ree] M[asons]," an appendix to *The Fruits of Grace, or Spiritual Thoughts of Two Lovers of Wisdom* by the prominent Rosicrucian Prince N. V. Repnin, one finds a somewhat different though no less instructive characterization of Masonry and its relationship to the Rosicrucians. According to this short text, the original Freemasons had been philosophers devoted solely to furthering the "common good," but as the order's membership grew its basic purpose eventually became perverted. Having lost touch with their fundamental goal, the Freemasons no longer gathered to share their knowledge with each other; instead, their meetings increasingly became devoted to hedonistic feasting. They began to admit to their ranks "the most debauched men without any selectivity and often for reasons of profit," and new members joined the lodges not out of an honest devotion to virtue and wisdom, but simply to satisfy their "empty curiosity." In response to this degradation, a select group established a "concealed brotherhood" isolated from the mass of Freemasons to carry on the order's original purpose.[76]

Repnin's text is significant for showing how the Rosicrucians viewed both themselves and the rest of the Masonic community, and for the light it sheds on the drive for distinction that so thoroughly shaped the world of Freemasonry. In addition, it supplies the proper context for understanding an oft-cited statement about Russian Freemasonry made by Novikov, who rose through the Masonic ranks himself to become one of the country's few Rosicrucians. Following his arrest in 1792, Novikov told his interrogator that decades earlier Russians active in "English Masonry almost played at it [that is, Freemasonry] as if it were a toy; they gathered, held initiations, had supper, and amused themselves; they accepted anyone indiscriminately, they talked a lot, but knew little." True Freemasons, on the other hand, whom Novikov (having left English Masonry) believed he had found upon joining the Rosicrucians, are "very small in number, and they do not try to collect members, [and] because of the great diffusion of false Masons at this time they must be highly secretive and remain in silence."[77] Novikov's testimony should not be read as a dispassionate and neutral assessment of Masonic practice, as past students generally have.[78] Rather, it provides yet another key to understanding the sense of moral and intellectual superiority that came with membership in the Freemasons and that in turn fed the development of ever more exclusive and elite groupings within their ranks.

That a movement like Freemasonry with its numerous degrees, grades, and titles found such resonance in eighteenth-century Russia is easily explained. The creation of the Table of Ranks in 1722 meant that for at least the next century and a half the acquisition of ranks was the chief avenue of personal and professional advancement for Russians seeking to better their place in society. The attainment of ranks became the raison d'être for generations of Russians, not only for persons of lower social origin but for members of the nobility as well.[79] Ranks were the ultimate prize after which all Russians chased. They were the primary indicators of social prestige. As the servant Polist aptly put matters in Ia. B. Kniazhnin's comedy *The Braggart* (1786):

> People have all gone mad about ranks.
> Tailors and joiners, merchants and cobblers—
> All share one faith and strive to become officers.
> And he who passes his dark life without a rank
> Does not seem a man to us at all.[80]

As an institutional structure, Freemasonry meshed well with Russian social and cultural norms that emphasized the importance of obtaining ranks and that equated rank with status. For Russians who failed to obtain a service rank, for those, in Kniazhnin's words, who literally did "not seem a man to us at all," or for those who felt that the possession of a rank still did not adequately express their own sense of personhood and dignity, the Masonic movement held out the promise of another set of comparable markers of status and importance. Here one could claim rank based on the nobility of one's soul, increasingly seen as the only true sign of nobility.[81] When Novikov sought to recruit A. T. Bolotov into the movement, he did so by trying to convince the skeptical Bolotov that in light of his personal qualities, his knowledge, and his virtue, he would quickly attain "the highest rank" among the Freemasons.[82] As Marc Raeff observed years ago, "in the moral and spiritual spheres Masonry offered a parallel or equivalent to the Table of Ranks in public service."[83] While this assessment is true in general terms, the distinction between the two spheres should not be drawn too sharply, for service to the public welfare occupied a prominent place in the

Masonic credo and by committing himself to his own personal enlightenment and moral betterment, the Mason saw himself as simultaneously working toward the improvement of society itself.

Freemasonry spoke to specific Russian needs and concerns in a variety of ways. On one level it provided an antidote to the perceived disorder of Russian society. Membership in the Masonic order signaled one's acceptance into a community of virtuous, honest, and enlightened men, and highlighted one's superiority over both the rabble and the evil flatterers at court. On another level Masonry's message of service to the public good, of molding oneself into a dedicated and trustworthy father, husband, citizen, and Christian, accorded well with the official ideology of the post-Petrine tsarist state. Finally, its fully developed system of degrees, which sought both to encourage and to reflect merit and achievement, was easily comprehended and readily adopted by a populace raised on Peter the Great's Table of Ranks. The elaborate hierarchy of Freemasonry gave Russian Masons like Prince Repnin and Novikov a way of expressing feelings about their own self-worth and personal dignity. It offered them, in short, an outlet for their vanity.

SECRECY

Along with what they reveal about the intricate relationship between the various Masonic degrees and systems, about the desire for exclusivity and distinction that so thoroughly shaped the Masonic world, the writings of Prince Repnin and the testimony of Novikov are also important for the clues they provide into the complex nature of secrecy within Freemasonry. Understanding of Masonic secrecy has long been shaped by widely held assumptions about the antagonistic relationship between the state and the lodges. According to this view, secrecy operated as a shield against the political authorities: given the totality of the absolutist state's claim to political power, the Masons could only exist behind a protective veil of secrecy. While this interpretation has some merit, it greatly simplifies the infinitely more varied and subtle functions and character of secrecy within Freemasonry.[84]

As the discussion of the Theoretical Degree and the Rosicrucians suggests, secrecy was used against other Masons as well as against the state or society. Clearly, the logic of secrecy reflected much more than a simple strategy for hiding from the political authorities, whose representatives, it is important to remember, formed an extensive part of the movement. In order to make sense of this logic, several questions must be addressed. First, what exactly was "secret" about Freemasonry? Second, where did the borders of secrecy lie, not only between the Masons and the profane world, but, equally important, among the Masons themselves? Third, what purpose did secrecy play in the construction of the Masonic world? And finally, what did the Masons perceive as comprising the movement's essential secrets?

With few exceptions, Russia's lodges met throughout the century free from state intervention. The absence of official interference, however, was not the result of ignorance of Freemasonry's existence within the empire. As early as 1747, Empress Elizabeth and A. I. Shuvalov, head of the Secret Chancery (Russia's political police), knew of the movement's inroads among the country's elite, and while they were rather suspicious of this foreign import, they took no steps to suppress it. An atmosphere of tolerance prevailed throughout Elizabeth's entire reign and on into that of Peter III (1761–1762), who was himself a Freemason and benefactor of St. Petersburg's Constancy lodge.[85] In the first part of her reign, Catherine II largely continued the policy of her predecessors. In the 1760s and 1770s, Russia's lodges met openly without provoking any repressive actions. The Masons showed no fear hosting even quite grand affairs, such as the concerts and soirées held in General P. I. Melissino's Modesty lodge in St. Petersburg or the large concert given at that city's Urania in March 1776, attended by nearly one hundred persons.[86] According to the Mason I. V. Lopukhin, not only did the Moscow police know about the local brothers' gatherings, they even sent units to ensure public order at the conclusion of especially large, festive meetings.[87] But by the end of the 1770s, Catherine had grown distrustful of at least some of the lodges. During a 1779 period of increased tension between Russia and Sweden, St. Petersburg Chief of Police P. V. Lopukhin was dispatched to investigate lodges directed by G. P. Gagarin. The lodges had recently shifted allegiance to the Grand Lodge of Sweden, and Lopukhin was to ascertain the exact nature of their links to the Swedish court.[88] It is interesting to note that during a visit to the imperial capital two years earlier, the Swedish king, Gustav III, attended two unusually large and sumptuous meetings arranged in his honor at the Apollo lodge that, according to one source, practically resembled "public festivals."[89]

Local officials acting on their own made similar attempts to gain information on the Masons' activities. Around the same time that Lopukhin visited lodges in St. Petersburg, I. V. Böber, then on his way from the capital to Reval to open the Three Poleaxes lodge, was detained by the Reval customs director, a Major Grenet, who had heard rumors that Böber was in the possession of Masonic secrets. His curiosity aroused, Grenet instructed his subordinates to search Böber's things, which did produce a cache of Masonic papers. Yet before Grenet had the opportunity to inspect them, Böber—far from intimidated by local officials—quickly enlisted the aid of powerful friends who demanded and immediately received the confiscated documents.[90] In Moscow during the mid-1780s, the governor-general Ia. A. Brius enlisted the young Mason V. P. Kochubei to spy on a group of local Masons whom Brius especially distrusted. Kochubei, who apparently informed Novikov that he had been charged to spy on him and his fellow Masons, was purposely raised into the higher degrees and allowed to attend all meetings so that he might report back to Brius on the harmless nature of local Masonic activity.[91] The cloud of mistrust hanging over these Moscow Masons cast its shadow upon the brothers active in St. Petersburg as well. In 1784, I. P. Elagin, apparently without direct pressure from Catherine or any other prominent authority, chose to cease temporarily the operations of his united lodges till the shadow passed.[92] This it appears to have done rather quickly, and by the next year Masons were once more flocking to the capital's lodges. In March 1785, for instance, a large crowd gathered in the Urania lodge—which had not even bothered to stop its meetings during these years—to listen to an oratorio composed by one of the brothers.[93]

The latter years of the century, then, did witness a growing mistrust toward the lodges on the part of the authorities, although this did not necessarily force the Masonic movement to hide its existence or to adopt a more pronounced form of secrecy. The case of Kochubei actually suggests the opposite: when confronted by an increasingly intrusive state, some Russian Freemasons chose to open their lodges and expose their entire affairs to full view. Secrecy did not develop as a response to political pressure but was a central feature of Freemasonry from its inception. Secrecy was less a response to the threat of state interference than an important factor in arousing the curiosity and suspicion of persons outside the movement, who in turn increasingly sought to open the Masons' activities to public scrutiny.

The diary of Aleksei Il'in, an ardent young Mason and Senate clerk in the mid-1770s who kept a thorough record of his daily affairs, sheds light on the boundaries and general character of Masonic secrecy. Il'in's diary suggests that Freemasonry was a common topic of conversation outside the lodges among both Masons and individuals who did not belong to the order. It also suggests that Russian Masons—at least during this period—were not particularly concerned with hiding their membership in the order; rather, one's status as a Mason seems to have been common knowledge among one's friends and occasionally even among one's broader circle of acquaintances. Despite general prohibitions against discussing Masonic matters outside the lodge, Il'in recorded doing just that on numerous occasions.[94]

During an outing with two friends to Moscow's vauxhall on 3 August 1775, for example, Il'in reported overhearing that in a few days he, along with his lodge brothers Prince G. P. Gagarin and A. B. Kurakin, would be initiated into the Master degree.[95] Il'in also met brother Masons and discussed lodge affairs while at work at the Moscow office of the Senate. On 3 June 1775, he was visited there by P. V. Chernov, a member of the St. Petersburg branch of the Equality lodge, who told Il'in all about its activities in the capital. Two other times that summer Il'in received in the Senate messengers from different lodges who had come to inform him of future meetings. On the latter occasion he also received congratulations on his forthcoming initiation into the Master degree from his lodge brother A. P.

Pronchishchev and Pronchishchev's wife, Anna Ivanovna.[96] After leaving the Senate that same day, Il'in visited the Pronchishchevs at their Moscow home where a certain Avdot'ia Petrovna (most likely Pronchishchev's sister), who perhaps had overheard them in conversation about the upcoming meeting, proceeded to poke fun at Freemasonry.[97]

At the end of 1775, Il'in moved to St. Petersburg where his days and nights were filled with an unending stream of sociable pursuits. There were constant trips to assemblies, to concerts at one of the city's music clubs, to balls, to the English Club where his brother, Petr, was a member, and to several Masonic lodges. As it had done in Moscow, Freemasonry continued to occupy a central place in the young clerk's life, as demonstrated by Il'in's decision to conclude his first full day in the capital with a visit to the Urania lodge, where, he laconically observed, "it was merry."[98] Here too Masonry was a topic of conversation among members of society, and Il'in's Masonic activities were common knowledge to his male and female acquaintances. Among these were V. G. Elagin—a director of the State Assignat Bank and a writer, not known to have been a Mason—and his wife, Anna Petrovna, in whose home Il'in lived and where he served as tutor to their children. Upon returning from the lodge one evening early in 1776, Il'in happened upon the mistress of the house who asked him what had transpired at his meeting. Very likely startled by this question, and the knowledge of his Masonic activities that it presupposed, he cryptically scribbled in his diary, "I did not respond to this and she came to believe that three years ago I was initiated. And Vladimir Grigor'evich also learned of this."[99]

The brief encounter between Il'in and Anna Petrovna reveals the primary boundary of Masonic secrecy. What was secret about the lodges was not their existence or necessarily their membership, but rather what transpired among the Masons inside them. This fact was explicitly stated in the Exegesis of the Statutes or Regulations of the Freemasons:

> You must especially bear in mind that law which you promised before the face of the heavens to observe with exactitude: the inviolable law of secrecy concerning our rites, ceremonies, signs, and manner of initiation. Tremble at the thought that this oath is any less holy than those which you swear in society. You were free when you took this vow, and you are no longer free to violate the secrets to which you are bound.[100]

The silence imposed on the Masons' activities, the curtain behind which they retired to carry out their meetings, was intended to lend them an aura of mystery and importance. By placing a veil over their actions, the Freemasons sought to mark their space—and, consequently, themselves—as sacred and distinct from the profane world.[101] Masonic secrets represented what Erving Goffman has described as "inside" secrets, the possession of which "marks an individual as being a member of a group and helps the group feel separate and different from those individuals who are not 'in the know.' Inside secrets give objective intellectual content to subjectively felt social distance."[102] For the Masons, who claimed to inhabit virtue's refuge and to be thoroughly engaged in its cultivation, secrecy provided a mechanism for strengthening their sense of exclusivity and superiority. It was a way of making virtue their own and themselves the sole proprietors of this precious commodity.[103] Lopukhin expressed this attitude in his memoirs:

> People especially decried the secrecy of the society and its assemblies. Why, they asked, must good deeds be done in secret? The answer to this is simple. Because in the gatherings of the so-called best people, or the public, not only does one not talk about, but it is in fact impossible to talk about God, virtue, eternity, the vanity of life, and about how depraved people are and how necessary it is for them to work toward their own improvement.[104]

The act of possessing something that was explicitly denied others bound the Masons together and enhanced their sense of constituting a separate and superior social body. Moreover, the Masons' ability to remain silent, to keep the secrets to which they were privy, was also perceived as an expression of this superiority. In other words, the capacity to keep still, to refrain from speech that could be considered harmful or dangerous, represented another mark of distinction since it attested to one's attainment in self-discipline, in working the rough stone.[105] Echoing this sentiment, one anonymous Russian Mason recorded in his notebook of Masonic teachings that "[t]he wise man always hides from others within himself. His eyes are open, but his lips are sealed."[106] The Mason's grim oath of allegiance to the brotherhood, sworn before he was allowed to see the light of Masonry, vividly conveys the great consequence accorded this form of self-mastery:

> I promise to be cautious and secretive, to suppress everything that is confided to me, and not to do or undertake anything that could reveal it. In case of even the most minor infringement against this obligation of mine, may my head be cut off, my heart, tongue, and entrails torn out and thrown into the sea's abyss, my body cremated and its ashes scattered in the wind.[107]

The graphic finality of this image shows again the attempt to construct Masonic space as the sole locus of virtue amid an endless expanse of vice. Here the contrast between the two spheres is underscored by exaggerating the danger posed by the profane world. Given the depravity and wickedness of the world, the Masons—as the only beacons of light and goodness—cannot tolerate the least indiscretion since to do so would be to risk extermination. Such a strategy for demarcating virtue mirrored the narrative

Assembly of Freemasons for the reception of Apprentices. With right hand on the Bible and left holding a compass to his breast, the candidate swears his oath to the order and pledges to keep its secrets or pay with his life.

technique common to that century's literature of sentiment in which, John Mullan writes, "virtue is most excited, and therefore most manifest, when threatened."[108]

As a group that wanted to be known for its superior virtue and enlightenment, the Masons could not take the demands for secrecy too far, because if they did no one would know anything about them. S. A. Tuchkov recorded in his memoirs how, as a young boy in Kiev, he listened to the officers visiting his father's home as they sang Masonic songs. These songs made such a strong impression on him that he copied them down and committed them to memory.[109] The way in which Il'in and his colleagues spoke so openly about Masonic affairs lends credence to Catherine the Great's reported assertion that the mystery of Freemasonry was similar to that of the theatrical secret whispered loudly enough by actors for everyone to hear.[110]

Although primarily directed toward the outside world, secrecy was never meant solely to separate Masons from non-Masons. Indeed, upon closer examination it is evident that secrecy was as common *inside* the lodge among the Freemasons themselves as it was at the lodge's outer doors. The use of secrecy among the brothers exhibited a preoccupation with creating hierarchies and levels of distinction similar to the Masonic development of higher degrees and competing systems. Just as the curtain of secrecy segregated the Masonic community from the profane world and imparted to its members a sense of exclusivity and superiority, so the boundaries of secrecy separating the numerous Masonic degrees played a similar role. While admission into the order marked the new Apprentice as morally superior to the uninitiated and, consequently, worthy of the knowledge of Freemasonry's laws, goals, and basic tenets, as a neophyte he had yet to earn the right to full disclosure of all its secrets and mysteries. In relation to fellow brothers of higher degrees, the Apprentice was purposefully held in a state of ignorance. Having earlier vowed not to divulge any knowledge to individuals outside the order, the new Fellow Craft Mason pledged to maintain the same vigilance in keeping all he was about to learn secret from his brothers in the Apprentice grade as well.[111] A similar pledge not to reveal secrets to the lower two degrees was also made by each new Master Mason:

> Do you promise, dear brother, with the renewal of the same punishment and the same oath that you swore as an Apprentice, to conceal inviolably from Fellow Craft Masons everything that you have already seen and that will be entrusted to you just as until now you have steadfastly hidden Fellow Craft secrets from Apprentices and Apprentice secrets from outsiders and the unenlightened? Do you promise not to divulge to anyone the least fact pertaining to Freemasonry of whatever grade known to you now or in the future?[112]

It is difficult to know how seriously these precepts for mutual secrecy among the members of the order—and especially of the same lodge—were followed. Once more Il'in's diary offers a clue for understanding how Masonic secrecy operated. A few weeks after being initiated into the Master degree, Il'in was visited at his Moscow home by L. V. Tred'iakovskii, a fellow member of Equality and Il'in's immediate superior in the Senate. Il'in recounts how during their conversation about the last meeting, "I began to tell him about the M[asonic] initiation. But immediately remembering that he is not a Mast[er], I stopped my story, and he did not suspect a thing."[113]

In Il'in's words, the subtle interplay of exclusion and secrecy at work in the lodges is evident. At the same time that Freemasonry sought to create among men such as Il'in and Tred'iakovskii a sense of belonging to a united community of virtuous and equal men, it also erected divisions among them, cultivating differences and fostering the creation of smaller secret worlds within their ranks. The reason for this practice was the notion that the degree of access to the order's secrets was directly proportional to the degree of one's personal development. As he rose from grade to grade the Mason proved himself ready and deserving of ever greater knowledge. Virtue and enlightenment—attainable only through self-examination and self-discipline—were tightly intertwined; only those who diligently and successfully worked the rough stone could ever hope to acquire knowledge of the powerful secrets located within the order.[114]

With increased access to the order's mysteries came a concomitant desire to make known to lesser brothers one's possession of these secrets. Much of the attraction behind keeping secrets derived from the pleasure felt in making others aware that one knew such secrets and in suggesting that one might be willing to share them. Such transactions clearly marked the unequal relationship between the participants and served to highlight where the greater power lay. Shortly after arriving in Russia in August 1775, Baron de Corberon, then a secretary to France's minister to the tsarist court, dined with a native Mason by the name of Izmailov. According to Corberon, the Russian officer

> told me, apropos of Masonry, that he was supposed to receive the Scottish grade. I gave him to understand that I was very advanced in such matters, that I had the power to impart my knowledge and to make anyone I wanted a Mason, as a simple matter of fact. I wish to use this [power] to make Izmailov my friend [and] to prevent him from developing a bad opinion of me. Moreover, this reputation of being able to accomplish wonders will make itself known, and this is a benefit with the women.[115]

Unfortunately, it is not known whether Corberon ever made good on this implied offer of assistance. (Nor is it known whether Corberon's Masonic reputation enhanced his status among the women of St. Petersburg.) It is clear that some advanced Masons selected brothers for initiation into the higher mysteries—even though these brothers lacked the proper degrees—only to deny them further admittance to their meetings. J. P. Wegelin wrote that upon joining the order in Moscow his superiors allowed him to skip the lower degrees altogether and exposed him almost immediately to the movement's great mysteries. Yet not long thereafter, Wegelin was purposely kept out of these higher-degree meetings and restricted to the order's lower levels. With great bitterness he related the story of his brother-in-law, a recently initiated Apprentice, who was invited to attend another Moscow lodge. To his dismay, the unwitting Apprentice found himself at a gathering of Scottish Masters, to which he was never again permitted to enter.[116] While it is possible that these incidences were the result of poor planning and communication, the tone of Wegelin's narrative suggests that they were intentional actions meant to reinforce Masonic hierarchy and ensure the loyalty of lower-degree Masons: lesser brothers' curiosity could be piqued, and their desire and envy incited, through brief glimpses into such unknown areas within the order. The superiors must have derived a pleasurable sense of their power by opening momentarily the door of their assemblies to the undeserving, only to shut it forcefully in their faces.

As the Mason climbed ever higher in the order and the veils that separated him from enlightenment were progressively withdrawn, he was able not only to see that which had previously been invisible, but he also learned to perceive that which he already knew in a different light. The various meanings ascribed to the traditional Masonic knock provide a telling example of this process. To the Apprentice, the knock—made by two short raps followed by a third, louder one—signified "the three punishments intended for David from God as mollification for his sins." Fellow Craft Masons interpreted the knock as referring to "[t]he two months of rest that Solomon gave to the chosen men after they had cut down the trees on Mount Lebanon" and "[t]he month of work for the same chosen men which was the third and most difficult."[117] For the lodge's Master Masons, the three knocks took on personal meaning. To them, the knocks were inextricably wed to the story of the murder of Hiram Abiff, the chief architect and builder of Solomon's Temple, who was struck down by three blows from three rebellious Fellow Craft Masons as they sought to force from him the secret Master's word. The story of Hiram's murder, which was ritually reenacted during the initiation into the Master degree, captured in the most dramatic terms how secrecy was used to strengthen the divisions within the lodge: as Masters, they alone were privy to the regrettable history of these mutinous brothers and through the shared knowledge of this episode were reminded of the divisions that separated them from the Fellow Craft brothers.[118] The message of caution toward the Fellow Craft Masons conveyed by this story was reinforced in a variety of ways, such as the three white roses on the Fellow Craft apron, which the Master Masons interpreted as signs of the "three pernicious blows which killed our most honorable Father and Grand Master."[119]

Although advancement from one degree to the next has so far been described as vertical movement (from bottom to top), Masonic discourse also depicted advancement as horizontal movement (from outside to inside). The map used to express these spatial relationships and to chart progression through space depicted the Temple of Solomon, the only perfect edifice ever known, built by divine wisdom but later destroyed, upon which the lodges were modeled and which the Masons strove to reconstruct. The temple was divided into three discrete spaces, one for each degree within which the brothers worked and learned the prescribed lessons of that grade before being allowed to cross the threshold leading inward. As a new member, and consequently nearest the profane world, the Apprentice worked "on the three thresholds of the temple and in its court"; next, the Fellow Craft worked "in the exterior part of the temple," that is, in the sanctuary; only the Master Mason worked "in the temple" itself.[120] The contrast between the proper place of the Master Masons and that of lower-degree brothers was also conveyed by the expression "Opening on Center," used in reference to gatherings of the third degree. While some have interpreted this to signify the Master Masons' rightful place at the center of the order (the Apprentice and Fellow Craft grades occupying its exterior and inner

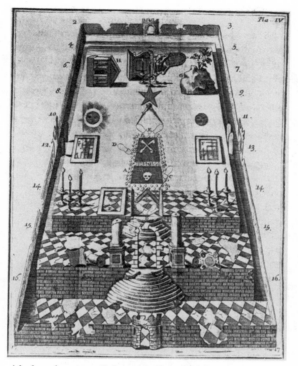

A Master Masons' lodge shown as Solomon's Temple. The Temple occupied an important place in Masonic cosmology as the perfect and most spectacular of structures. It served as a model for the lodges, and the legends surrounding its construction formed the basis for the initiation ritual into the Master Mason degree. The Temple's three sections are mirrored by the order's three levels: as the brother progressed to the next degree, he moved ever upward and inward toward the Temple's holy of holies.

circles respectively), others have interpreted the expression as referring to the circle formed by the Master Masons at the beginning of their meetings, in which all stood at an equal distance from the circle's center.[121] In either case, the expression denotes the connection made between higher levels and interiority, and lower levels and exteriority.

Thus, the higher one ascended within the Masonic hierarchy the deeper one penetrated into the sacred space of Solomon's Temple. With the progressive movement upward and inward came initiation into ever smaller, more select bodies understood to lie increasingly near the temple's (and the order's) center where Freemasonry's essential mysteries dwelled. Yet, the center toward which every Mason strove turned out to be quite elusive, exhibiting a tendency to shift out of sight at the last moment. Having arrived at what they expected to be its innermost space, Master Masons in the

Rising Sun lodge, for example, learned that the Temple of Solomon actually consisted of nine separate chambers, not three, and that they had entered into only the eighth, penultimate one. The holy of holies, it turned out, still lay ahead.[122] These brothers, who had come to the end of St. John's Masonry, now received intimations that only by proceeding further into the order and joining the higher degrees would the ultimate meaning "of our praiseworthy undertakings in these holy works" be made fully known to them.[123] As one Russian Mason noted in his reminiscences, "The secret is to be found only in the heart of the order."[124]

The Scottish degrees usually represented the next step toward that heart for brothers who had come to the end of St. John's Masonry. The border separating these degrees was impressed upon initiates into Scottish Masonry, who swore to keep secret "not only from outsiders but also from brothers of St. John's Masonry" everything henceforth entrusted to them.[125] Andrew's Masons reconfigured the temple to reserve its innermost spaces for themselves. One set of texts divided the lodge into four distinct zones—"the square, threshold, holy place, and holy of holies"[126]—the last of which belonged to the Scottish Masters (the highest degree), the initial three Masonic degrees having been removed to the more exterior regions of the temple.[127] In the map of the order drawn by the Rosicrucians, who occupied its highest ranks, the brothers of St. John's Masonry were displaced to the "threshold" of the temple, within the "outer wall of silence" dividing Freemasonry from the "unenlightened," yet outside the "inner wall" that separated them from the internal order (that is, the Rosicrucians).[128] "Freemasonry," one text popular with Russia's Rosicrucians asserted, "is the forecourt of the temple, whose secret entrance only the worthy Freemason can discern and open."[129]

The Rosicrucians were not the only ones to relegate their lesser brothers to the movement's periphery, claiming for themselves the symbolic center invisible to all but the most worthy. Deeply immured within the figurative temple of Freemasonry, the highest levels of several Masonic systems were hidden from fellow brothers of lower degrees. In St. Petersburg in 1775, for instance, Melissino founded his own eponymous system comprising seven degrees. The highest degree, the Magnus Sacerdos Templariorum, was bestowed upon only a select number of brothers organized into a secret committee called the "conclave," unknown to the rest of the membership.[130] Within the Swedish system introduced to Russia by Prince Kurakin in the late 1770s, the Grand National lodge functioned as the visible governing body. The actual locus of authority, however, was a smaller body called the "chapter," whose members were drawn from the Grand National lodge and whose existence remained secret to the rest of the system's brothers.[131] The Second Elagin Union of English lodges formed in 1786 also possessed a secret chapter, which operated in the shadows of the nominal supreme body (the lodge Modesty). Named the "High Chapter" in 1790, this internal structure admitted only "the most well-trained and deserving members of the brotherhood," upon whom was conferred the eighth degree, the remainder of the union's lodges being limited to working in seven.[132]

Of all the systems operating in Russia, the Rosicrucians clearly had the most developed form of internal secrecy. Atop the regular St. John's lodges and Scottish Masonry, they erected several secretive bodies, beginning with the Theoretical Degree, that operated unbeknownst to brothers of lower levels. The Rosicrucians in turn remained hidden from the Theoretical Brothers, from the lodges subordinate to them, and from the broader Masonic community. Being secret and hidden was of central importance to the Rosicrucian identity. As N. N. Trubetskoi, a leading figure in the order, opined in a letter from September 1783, "true" Rosicrucians were by definition "secret" and "unknown brothers," an identity reinforced through the adoption of cryptic names for use within the order.[133] Trubetskoi's understanding of the Rosicrucians was in keeping with the first discussions of the group from the early seventeenth century when the notion of a secret, ancient brotherhood invisible to most of humanity first surfaced in Europe.[134]

In certain instances, while the specifics of these highest bodies remained unknown (that is, who their members were, when and where they met, what sort of structure they possessed), their presence and authority were made manifest to their subordinates. Baron de Corberon had been informed of the existence of Melissino's conclave and even of its mysterious alchemical experiments almost a full year before being admitted to it in 1777. Whereas a year earlier, Corberon had enjoyed superior knowledge of Masonry's mysteries in his dealings with the Russian Izmailov, he now found the shoe on the other foot and wrote how pleased he would be when he was finally admitted to this exclusive body.[135] Initiates into the Theoretical Degree learned that above them a "higher internal brotherhood" existed, to which they could eventually be presented but only after having demonstrated the prerequisite personal development. The composition of this brotherhood, its location, and even the basis of its authority, however, were not divulged.[136] Not surprisingly, such an arrangement did not fail to elicit suspicion among members of the Theoretical Degree. A speech harshly reproving a group of Theoretical Brothers in Orel for their distrust of the unnamed "superiors" and the "invisible hand" used to guide them shows that the authority claimed by hidden and secretive groups did not always go unchallenged and could lead to dissension within the order itself.[137]

There existed yet another side to Masonic secrecy that was not officially sanctioned and whose effects are much more difficult to trace. This side had to do with the subtle stratagems the brothers employed against each other—having nothing to do with accepted laws and procedures—that could be directed against Masons at all levels of the Masonic hierarchy. The correspondence between Prince A. B. Kurakin and several brothers in Stockholm concerning the establishment of Swedish Freemasonry in Russia offers telling insights into this murkier facet of secrecy. Having been initiated into the mysteries during his recent trip to Sweden, Kurakin wrote on 8 February 1777 to Charles, duke of Sudermania, brother of King Gustav III, and the head of Swedish Masonry, about how he was surreptitiously preparing Russian Masons for the arrival of the new system. Most shocking in this letter was Kurakin's plan to manipulate none other than Elagin, the most respected and powerful Mason in all Russia. Given Elagin's standing in the order, Kurakin planned to make sure that the brothers elected Elagin Grand National Master: "But he will be elevated to that important post only under very significant conditions: he will have for himself the ceremonial aspects of authority, but I will have the true power."[138] Kurakin's words exposed the intricacies of Masonic secrecy that were even more complex than the maze of degrees, systems, and hidden bodies suggests. Not only were there invisible hands at work, whose existence one knew of or at least suspected, but there were also invisible hands that remained completely unknown.

Secrecy served to establish or to reinforce sharply delineated regions of visibility and invisibility. Each degree and group within the order occupied its own space inside the temple, and these spaces were clearly walled off from each other by opaque barriers. No one could see into the temple's next chamber, nor even be sure what other chambers existed. The higher the Mason rose in the hierarchy, the greater became his knowledge of the temple—of the rooms that made it up, their occupants, and the specific knowledge revealed in each one. Individuals outside the movement remained ignorant of what went on inside the secret world of the lodges. Their range of sight extended only to the outer doors of the lodge and no further. The Freemason, on the other hand, had the knowledge of two distinct worlds operating before him; both were visible and, hence, knowable to him.

Of course, it should be pointed out that the profane world was not undifferentiated, flat, and uniform; it too had sites bounded against the curious gaze of outsiders. The Russian social landscape was divided into myriad discrete zones, each with its own borders that allowed for varying degrees of access, exclusivity, and transparency. One might note the preference among the haut monde for private theaters whose atmosphere of heightened intimacy and privacy stood in marked contrast to the public theaters with their general accessibility and "less refined" patrons. Indeed, the relationship between private and public theaters is homologous to that between the Rosicrucian circles and the Masonic lodges: just as the new private theaters provided Russia's aristocrats with an alternative to the public theaters frequented by the hoi polloi, highly secretive and selective bodies such as the Rosicrucians provided men like Prince Repnin with an alternative to the more popular and common Masonic lodges. But clearly the primary example of a discrete sphere shrouded in indiscernibility was the tsarist court—the principal animating center for all of Russian society—with its competing patronage networks, cliques and cabals, ruthless Byzantine intrigue, and complex unwritten rules for social interaction. In a manner strikingly similar to the practices discerned among the Masons, Catherine the Great established at court the "Little Hermitage," a select company of favorites, ladies-in-waiting, and courtiers that met, frequently in masque, several times a week for dances and games of all sorts. Despite the intimacy and exclusivity afforded by this company where, according to the Frenchman Charles François Philibert Masson, "the greatest privacy prevailed," Catherine went on to create "another assembly, more confined and more mysterious, which were [sic] called the Little Society." It was amid this smallest of societies, limited to no more than a handful of the empress's closest intimates, that "this Cybele of the north celebrated her most secret mysteries," mysteries so powerful, so rare that even Masson, who claimed to have been admitted to this coterie and initiated into its secrets, refused to divulge them.[139]

In its veils of secrecy, its baroque hierarchies of degree and title, and its ever more select bodies shrouded in mystery, Freemasonry mirrored court society, forming a species of "counter-image of the official world."[140] As most Masons found themselves excluded from the few theaters of the great nobles and the world at court, the lodges provided a secretive space in which they might cultivate their own sense of exclusivity and separateness. For those Freemasons drawn from the uppermost strata of the social order, the lodges furnished yet one more arena in which to rehearse their predominance and status, albeit in a decidedly different key.

The Masons, then, inhabited a veiled counterworld set up amid a social reality that, while not without its own spheres of exclusion and concealment, remained immanently more transparent and knowable than that of the lodges. At home in both spheres, Masons were privy to two separate realities. In a sense, the world presented more of itself to them, made more of itself visible to Masons than to non-Masons, and this greater range of vision increased as the Mason climbed the institution's ladder. With each initiation into a higher level, the limits of sight were pushed back ever further. One saw more of the world, light having been cast upon its dark recesses and shadowy niches. This enhanced field of vision pertained not simply to the mysteries of Freemasonry or to the hidden nature of the physical world, but to one's fellow human beings as well. Expropriating the power of the "all-seeing eye," a chief Masonic symbol intended as a sign of the omnipresence of the supreme being, higher-degree Masons professed the ability to pierce visually their lodge brothers' outer shells and to discern their deepest thoughts and the movements of their hearts.[141] Masons in the lower degrees were frequently admonished to open themselves up to their superiors, to make known voluntarily their inner states, since there was no possibility of hiding them. In a speech to the Theoretical Brothers in Orel, Zakhar Karneev announced that "[e]verything, even the slightest movement of our hearts is known, discernible and far more visible to the Holy Order and the superiors stationed above us than to ourselves."[142]

As Karneev's words suggest, questions of sight and visibility were tightly bound up with the dynamics of power and discipline at work in the lodges. To see was to possess information, and one's field of vision was directly proportional to one's place in the Masonic hierarchy. Sight extended in only one direction: downward, both from the heights of the Masonic movement onto those of lower grades and from the lower grades onto the profane world. The summit was often controlled by unseen figures like the Rosicrucians, who remained invisible and unidentified to everyone outside and inside the Masonic brotherhood.[143] Within their own ranks, the Rosicrucians were under the authority of the "unknown superiors" whose identity and physical location were not divulged to anyone, not even to other Rosicrucians.[144] Their bylaws provided numerous justifications for the "wisely sanctioned concealment of all the O[rder's] superiors," claiming that it

> a) . . . protects them from . . . the thirst for glory and the lust for power that usually through self-love and pride are able to seduce and entrap imperceptibly many people, and no less from all impudent acts of the lesser brothers detrimental to the general order since no one can know who they are.
>
> b) inhibits personal hatred and envy among all, and especially toward the superiors.
>
> c) . . . keeps them [that is, the lower brothers] in a state of obedience appropriate to their position . . .
>
> d) counteracts treachery and greed as well as the vindictiveness of perfidious brothers . . .
>
> f) allows for the good of the Order the sublime superiors to visit lower gatherings, to uncover irregularities by various means, and to issue the most useful commands so that they may be corrected.[145]

Through their invisibility the supreme authorities transformed themselves into what Georg Simmel has described as an "intangible power, whose limits cannot be traced, and which can nowhere be seen, but must, for this reason, be suspected everywhere."[146] And it was more than simply the invisibility of the Rosicrucians' leaders that gave the brothers good reason to suspect their omnipresence. According to the order's charter, the superiors

> see everything that the brothers do on earth, how they behave, where and with whom they socialize. They see if they are in danger, if they live in fear of the

Lord, or if they drink heavily, overeat, whore, gamble, or commit any similar sins; they see in what sort of circles they move; . . . if they avoid all luxury and extravagance. Do they fulfill their duty as fathers, do they set a good example to their wives and children, do they give themselves fully to them and devote themselves to the Lord; . . . do they visit the sick, the confined, do they clothe the naked, comfort with all their heart the wretched and the abandoned?[147]

When considering the Rosicrucians' supreme visual powers it is important to recall that this unlimited ability to see was as much a function of their position at the order's center as at the top of the Masonic hierarchy. The Rosicrucians claimed for themselves the sanctum sanctorum, the innermost chamber of Solomon's Temple, and assigned in descending fashion the remaining spaces of the temple progressively nearer the outer world to brothers of lower degrees. The preoccupation with inner and outer states (the notions of inside and outside themselves) was not specific solely to the Rosicrucians but resonated throughout Masonic discourse, thoroughly shaping the entire Masonic worldview. In this sense, Freemasonry offered its own variations on inherited themes pertaining to the opposition of inside to outside, interiority to exteriority, truth to falsity, and reality to appearance. A few of these oppositions have already been touched upon. One was the distinction frequently drawn between morals—reflecting internal states and essences—and manners, often perceived either as a deceptive mask for moral depravity or as an accurate expression of moral goodness. Another was the opposition of the lodge to the outside world; here the use of secrecy as a veil marked the lodge as an invisible inner world in contrast to the visible outer world. The entire Masonic movement was portrayed as built upon internal essences (virtue and morals) in explicit contrast to society, which was depicted as built upon external appearances (wealth and rank). Since he was not in society, in the lodge each man was judged (if only in theory) according to his inner worth and not his outer status.[148]

During his induction into the order, the candidate was already taught to think in terms of such distinctions. Blindfolded and waiting in the black chamber, he was instructed to go inside himself, to look into his heart, and to acquaint himself with the workings of his inner self. Here he would find the truth, for it was not in empty words or in the outside world that truth resided, but inside him, in his true internal essence.[149] The opposition of inside to outside was central to the construction of Masonic identity. In contrast to the Mason's depth, timelessness, and hence, truth of character stood the "external man," the Mason's constituting other, who remained blinded by and enslaved to the fleeting pleasures of this world, to everything transient, and ultimately artificial, unreal, false.[150] These themes were vividly conveyed by the speeches of S. I. Gamaleia delivered to the brothers of Moscow's Deucalion lodge in the early 1780s. Gamaleia singled out for condemnation man's insatiable thirst for fine clothes, large houses, and high ranks, objects that he claimed belong to the world of exteriority and that were most often nothing more than masks intended to cover inner vices.[151]

In Gamaleia's orations, Masonic discourse was used as a general critique of social reality. Appearances were frequently deceiving, and individuals of great wealth and position could prove upon closer inspection to be far from deserving of such privileges. Masonic texts often pointed out the danger of judging persons "by outward appearance" and reiterated that man's greater value lay below the surface.[152] Nevertheless, it would be a mistake to see in this language of the internal man a critique of Russia's political and social order. Palaces and positions of authority were not inherently wrong, but only if they constituted the primary objectives of one's actions and not the fair rewards for faithful service. As mentioned in chapter 1, the Masons strove to outdo all others in fulfilling their official duties. Each new Mason was instructed to carry out the laws and commands of state and church not only by their "letter" *(po edinoi tol'ko naruzhnosti)*, but by their "spirit" as well.[153] Whether or not his deeds won him recognition in the "empty celebrations of the common crowd," they would more importantly be recorded "in the tender voice of his conscience."[154]

The same semantic pairings associated with interiority and exteriority that organized Masonic perceptions of the physical body also structured the images of Freemasonry's institutional body. Some of these images, such as those constructed with the model of Solomon's Temple, have already been discussed, but others are equally important for the light they shed on Masonic cosmology. The Rosicrucians, for instance, described themselves as the "supreme internal brotherhood" or the "internal Order." Beyond their outer border lay the Theoretical Degree—the midpoint or "useful mediation"—between Rosicrucianism and lower-degree Masonry situated on the boundary separating the order from the outside world.[155] The language of interiority was also exploited in the internecine struggles for institutional predominance among Russia's Masons. In a letter from the early 1780s to Ferdinand, duke of Brunswick (one of Germany's leading Freemasons), a group of Moscow Freemasons bemoaned the movement's present state in Russia for having fallen under the sway of "templarism," characterized by (in their words) a strong preference for and preoccupation with "[s]plendid ceremonies . . . crosses, rings, [and] cloaks." The reason for this regrettable development, they continued, was that Russia's lodges were dominated by the "high" and "wealthy nobility" whose members, having been raised in a "very sensual manner," were used to spending their lives either "ornamented with such signs of honor" or "seeking nothing so greedily as their attainment."[156] In this view, Russia's leading Masonic system represented a corruption of the movement's original meaning and intent. Having abandoned the chief task of refashioning the inner self, of replacing vice with virtue, the brothers in these lodges supposedly satisfied themselves with simply decorating their exteriors, covering themselves "from head to toe . . . in chivalric orders."[157]

Swedish Freemasonry, which had firmly established itself in Russia by then, was the main target of these criticisms, and it was in the hope of freeing themselves from Swedish control that the Moscow Masons wrote to the duke. Their desire to institute Russian Freemasonry on a more independent

Drawing from a Rosicrucian manuscript. The German text reads: "Oh, mankind, know thyself. In you the treasure of all wisdom lies hidden." While self-knowledge was the foundation for all Freemasonry, the plumbing of the self's inner depths acquired much greater significance in the order's higher levels. The quest for spiritual truth led some Russian Masons to establish contact with a group of Rosicrucians centered around the future Prussian king Frederick William II claiming to know the lone path to true enlightenment.

national basis led the Masons, ironically, to place themselves in the hands of yet another group of foreigners (this time the Germans) who soon thereafter designated Russia as constituting the order's VIII province.[158] The leaders of this new branch were none other than the Rosicrucians, whose ranks included Novikov, J. G. Schwarz, and the brothers Iu. N. and N. N. Trubetskoi. In their efforts to discredit Swedish Masonry and thereby win over Russia's lodges to their new alliance, the contrast between inside and outside appeared once again.

In several letters to A. A. Rzhevskii, whom the Russian Rosicrucians were seeking to recruit as their St. Petersburg head, N. N. Trubetskoi wrote that the Rosicrucians now represented the "center point" of the entire order. According to Trubetskoi, the "hand of God" had led them to the true "internal Order" that was to be found only among a highly select group of brothers in Europe, where "truth shines in its radiance without hieroglyphs." Here amid the Rosicrucians, he added, they had finally located what all Russian Masons had been searching for—"the single purest and brightest source" at the very heart of the order. The brothers of Swedish Masonry, on the other hand, remained on the outside, segregated from those who constituted the order's central core. Their exterior position was mirrored in the Swedish Masons' teachings; removed from and hence blind to truth's radiance, they had nothing to offer but empty figures and signs—mere hieroglyphs void of any inner essence or truth. Absent any substance and interiority, Swedish Masonry, for the Rosicrucians, was nothing but form, a meaningless play of mute surfaces and exteriors.[159]

As Trubetskoi's letters show, claims about spatial precedence were basically claims about knowledge. Truth and wisdom were held to be anathema to surfaces and exteriors. They naturally resided inside things and, more specifically, at their very center. Thus, when asked during their rituals "where does Wisdom reside?" the Rosicrucians answered: "In the light's center point."[160] And as the self-proclaimed inhabitants of this central space, they professed to be the sole possessors of wisdom itself.

The connections among these related themes of the body, the Masonic institution, and the order's teachings are neatly traced in an anonymous eighteenth-century manuscript comprising various short writings on the movement. In this text, which corresponds to previously discussed models, the order's institutional structure is divided into three constituent parts: parabolic, theoretical, and practical. Each of the three classes develops a distinct part of the individual—the physical body, the soul, and finally the spirit—that displays movement from exterior surfaces toward the innermost recesses of the self. Not only do the three specific sites for instruction move progressively inward, but the corresponding educational means to be used there also express an increasing degree of interiority. Whereas in the first group, "hieroglyphs, allegories, and symbols" (that is, external forms)

CHART 1

Parabolic St. John's Degrees	Andrew's and Theoretical Degrees	and the supreme internal degrees
In these degrees the letter is presented in the form of a hieroglyph	here the meaning of the letter	and there the spirit enlightening the meaning and animating the letter
morality belongs here	here the heart feeds on love	and there is the school of wisdom
here one acquires knowledge of man	here of nature	and there of God
here external natural means are used	here spiritual means	and there Divine

constitute the appropriate tools, in the third class, brothers progress beyond mere signs and discuss matters in their "essence and authenticity."[161] The reason for these distinctions in method and subject is a function of the degree to which one has developed the internal self. Thus, those in the lower levels remain "sensual and external persons" for whom the study of outer forms and signs is most appropriate; only after having adequately developed one's inner self can one move on to a study of the deeper inner truths communicated by these signs and symbols.[162]

Another contemporary manuscript communicates an almost identical message in the form of a chart entitled "The Course of the Order's Teaching" (see Chart 1).[163]

It is difficult to say exactly what these inner truths, these secrets and mysteries that supposedly lay at the center of Freemasonry, were. Initiation into the lodge brought with it knowledge of Freemasonry's ceremonies, rites, and laws, all of which were considered secret and were not to be divulged to anyone outside the order. Such were the secrets that Aleksei Il'in withheld in St. Petersburg from his master and mistress, Vladimir and Anna Elagin, who knew something of his Masonic affiliation but nothing of the movement's rites and rituals. The animating principles rooted in the texts and practices that structured Masonic life, both ordering the Masons' activities inside the lodge and offering specific guidelines and standards for behavior outside it, formed some of the order's fundamental secrets intended for the brothers' sole usage and benefit. If properly observed, the knowledge contained in them would show each brother the proper path to virtue and enlightenment, a path that, the Masons averred, few individuals sought and even fewer knew.

Still, however important the secret knowledge contained in their statutes and communicated through their exercises, it did not constitute Masonry's lone or even primary secret. There existed, or so they claimed, a greater and more potent secret buried deep within the movement, a fact each member was apprised of during his initiation:

> [W]e must hide our secrets [*tainosti*], that is our rites, form of government, etc. I say nothing about [our] sacraments [*o tainstvakh*], for no one can divulge them to anyone; they, as sacraments, are secretly communicated by those who possess the authority and the power to communicate them.[164]

The Masonic order, then, possessed two classes of secrets. The first of these became partially known to each brother upon admission, and was progressively revealed as he rose through the ranks; the second, while translated above as "sacraments" but perhaps better understood as "mysteries,"[165] was to remain unknown until such time as certain undisclosed persons of authority "secretly" communicated it to him. The notion that such an immensely powerful mystery formed the invisible axis around which the whole Masonic movement rotated, and whose existence supposedly provided Freemasonry's raison d'être, played a major role in the order's cosmology. Even before learning of the two types of secrets, initiates were told that chief among Masonry's three goals was "the preservation and the handing over to [our] descendants of a certain important mystery that has come down to us from the most ancient times, even from the very first man, and upon which may depend the fate of the entire human race, until such time as for the good of humanity God favors that it be made known to the whole world."[166] Wegelin attested to the awesome power of this mystery. Initiated directly into the higher degrees in obvious disregard for the established rules, he wrote, "I was allowed to look too deeply into the mysteries of the order . . . I was filled with the light of Masonry to such a degree of satiety, even to the loss of consciousness, that to this day I don't know how I was able to remain within Freemasonry."[167]

To attempt any thorough answer to the question of what precisely this mystery might have been is not possible for the simple reason that none of the Freemasons themselves could agree on what it was. Since it went unnamed—alluded to, pointed at, discussed, yet never revealed—the mystery within Freemasonry could be taken to be different things by different Masons. And since the mystery had to exist, given its prominence in Masonic lore, there arose an irrepressible impulse to fill the vacuum created by the many references to it, an impulse that, in turn, spurred varying understandings of what constituted the esoteric mysteries at Masonry's core. To Melissino and the members of his conclave, alchemy appears to have formed the basis of the essential mysteries. To the Moscow Masons sur-

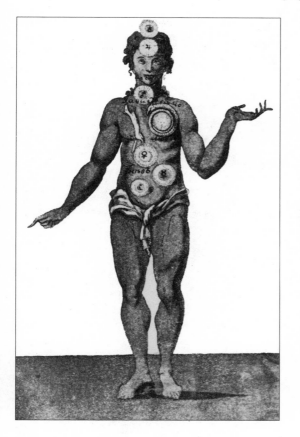

Symbolic depiction of man based on Jacob Boehme. The Rosicrucians' beliefs were largely based on the spiritual alchemy of the German mystic Jacob Boehme (1575–1624). Boehme wrote that after the Fall, man lost his initial harmony and descended into a state of "turba," or confusion, in which the earlier unity of the Three Principles and the Seven Qualities yielded to the chaos of the four elements. This image depicts the elements' dominion over mankind: fire is in his heart; water is in his liver; earth is in his lung; and air is in his bladder.

rounding Schwarz and Novikov, on the other hand, alchemy represented simply one strand of the mysteries to which they added the legend of the Rosicrucians, the writings of the German mystic Jacob Boehme (1575–1624), and the austere pietism of Johannes Arndt.[168] But the most ambitious (if not grandiose) attempt to define Freemasonry's mysteries clearly belonged to Elagin who, in his "Teachings of Ancient Secular and Spiritual Wisdom, or the Science of the Freemasons," tried to synthesize into one single work the prime truths of Zoroastrianism, the Greek and Roman classics, the Old and New Testaments, the Talmud, the Hermetica, the Cabala, and the writings of Robert Fludd.[169]

In the end it might be fair to say that the final mysteries of Freemasonry turned out to be whatever its leading representatives decided they were, and this, in turn, hinged largely on the intellectual fashions of the day. The interest in various currents of mystical and occult thought evinced by the Masons was a reflection of a broader transformation in the European intellectual world in general, as demonstrated by the popularity of the Viennese

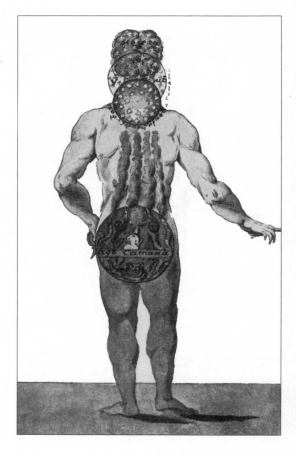

Symbolic depiction of man based on Jacob Boehme. Despite mankind's sorry state, Boehme believed that the lost balance of the Three Principles could be restored by expelling the bestial instincts that had gained control over it. This image, entitled "Practice—Man Perfected and Enlightened by the Divine Being," shows the principles once more in harmony. The light principle, represented by the celestial circle, comes into union with the dark principle, represented by the "Satan's Hell" of the buttocks releasing its stench, to produce the world of sense, illustrated by the top circle.

physician Franz Mesmer's Parisian séances, the wonderworks of Cagliostro, or the writings of Johann Georg Hamann in the later decades of the century. Indeed, after leaving St. Petersburg and the alchemical experiments of Melissino's conclave, Baron de Corberon returned to Paris where he joined the mesmerist Society of Universal Harmony and frequently visited Cagliostro at the home of Cardinal de Rohan.[170] Whether or not Corberon or any of Russia's Freemasons truly believed they had gained full insight into the order's eternal secrets remains unknown. Achieving knowledge of the order's "great mystery," I. P. Turgenev observed in the 1790s, was contingent upon the moral improvement of one's character to such a degree that one made oneself as close to perfect as is humanly possible.[171] And who could ever honestly claim that distinction?

. . .

*F*reemasonry represented a broad-based attempt to create a new symbolic order, one defined by its distinctness and separateness from the general social landscape, which the Masons equated with vice, corruption, and ignorance. In this attempt, Russia's Freemasons made use of what Hayden White has called "the technique of ostensive self-definition by negation," whereby the pronounced focus on the supposed sin and depravity of others is meant to trace the space of virtue and purity occupied by oneself.[172] The social reality created in the lodge was based upon norms and values different from those believed to govern society as a whole; indeed, the lodge was intended to be a place where prevailing societal norms were inverted, reversed: fellowship took the place of rivalry, love that of enmity, virtue that of vice, enlightenment that of darkness, substance that of appearance. Still, Freemasonry was the creation of men who inhabited the same world they so earnestly sought to distance themselves from, and as a result, the new world could not help but mirror on some level the larger social order of which it formed a part. Here one need only call attention to the Masons' heightened sensitivity to gradations of status and their predilection for intricate and clearly demarcated hierarchies of rank—attitudes and inclinations that were highly reflective of Russian society in general.

Yet, despite the deep structural affinities between the Masonic movement and society, it must be stressed that to the Freemasons, as well as to their critics (to be discussed in the next chapter), the lodges constituted a truly unique social space, whose structures and practices set it apart from all others. To its members, the lodge was located simultaneously at society's symbolic center and at its periphery. In proclaiming the lodge the lone refuge of virtue and themselves its sole protectors, not to mention its very personification, the Masons were attempting to remap Russia's moral geography by portraying Freemasonry as society's animating center.[173] But the manner in which the reconfiguration of the symbolic landscape was conducted—the exaggerated pronouncements of their isolation from the dominant social institutions, the construction of an opaque wall of secrecy intended to mark the division separating themselves from the rest of society, and the readiness to adopt the status of the besieged outsider—served in effect to displace the Masons to society's margins.

There was nothing accidental about this strategy and its outcome, however. Russian Masons' efforts to align themselves with the symbolic peripheries of society were rooted in the fact that these marginal areas are traditionally thought to be imbued with extraordinary powers. It is precisely to such liminal groups perceived as occupying the social fringes that one attributes special qualities of sacredness and moral superiority. But margins, as Russia's Freemasons would discover, possess a paradoxical and problematic nature. At the same time that they, and the groups that inhabit them, are suffused with elemental, life-giving powers, the margins of society are also invariably seen as dangerous places, ones frequented by the wicked and nefarious.[174]

The Image of the Mason

Who doesn't believe dreams, / Magicians or sorcerers
That there's a wood-goblin in the forest, / And a hob-goblin in the house,
That when salt spills, / When a cock crows,
That when a hen sings, / That when a cat starts to scratch,
Great misfortune, / Calamity and horrors ought be expected;
If a Papist walks by, / A Freemason approaches,
Raise your eyes to heaven, / And then you won't fall on your face.

—Ivan Barkov, "The Creed of Vaniushka Danylich"

The image of the Freemason played heavily on the Russian mind of the eighteenth century. One encountered him in all manner of contexts and in all manner of settings. Masons appeared in sermons and satires, in songs and speeches; they appeared in plays, poems, and parodies, in letters, articles, and advertisements, in reviews and reports, in official orders and interrogations. Masonry was denounced in churches and mocked on stages; it was debated in the gatherings of polite society and discussed at court; it was gossiped about in markets; it was belittled in the press and exposed in pamphlets and books sold in the country's bookstores. Within the brief span of a few decades, Freemasonry went from being utterly unknown and unheard of to being powerfully resonant both among Russia's public and among the illiterate masses in the urban centers and the isolated villages of Russia. By 1800, the image of the Freemason, having acquired over the years layer upon layer of meaning—not without some disharmony and obvious contradictions between them—had come to haunt the Russian imagination. For while Freemasonry could, and did, represent different things to different people, to the vast majority of Russians it invariably elicited two responses: suspicion and fear.

Russian views of Freemasonry can be divided into three general categories corresponding to three separate chronological periods, each period representing the addition of another perception of the order that did not necessarily efface the existing perceptions but coexisted with them, highlighting certain features, displacing others to the background. While this scheme is at base artificial, reducing the chaotic mass of images to an overly neat pattern, it nonetheless captures the principal ways Russians viewed the Masonic movement and shows how these impressions changed over time. It also makes clearer Masonry's place in society by shedding light on how those outside the order perceived it, explaining exactly what it was about the lodges and their members that so disturbed many Russians and provoked them to words of condemnation. In order to defend the lodges in the eyes of the public, the Freemasons responded to critics by writing their own accounts of Masonic life. Tracing the interplay between these opposing forces illuminates the public debate over Freemasonry and provides insight into the larger question of the role and dynamics of public opinion itself in eighteenth-century Russia.

SATAN'S SERVANTS

Early attitudes toward Freemasonry in Russia are difficult to establish with any great specificity. What is known beyond doubt is that the movement attracted attention soon after its arrival, and that this attention was largely negative. Although the evidence is sketchy, during the reign of Elizabeth knowledge of the lodges' existence seems to have spread beyond polite society into lower urban social strata where Freemasonry was generally regarded as a thoroughly illegal and ungodly affair.[1] According to one source, as early as the late 1750s rumors of a gathering of Masons in St. Petersburg were able to set off "panicky fear" throughout society.[2] In his memoirs, the poet Gavriil Derzhavin recounted that in the early 1760s, as a young man full of wanderlust and bored with life in Moscow, he approached I. I. Shuvalov, then head of Moscow University, hoping to be taken along on Shuvalov's upcoming travels abroad. Derzhavin's hopes were quickly dashed when word of his intentions reached his aunt, Fekla Savishna Bludova, who promptly forbade him to meet Shuvalov. An old-fashioned woman steeped in the ways of traditional Russia, Fekla Bludova was not about to let her nephew associate with a man many believed to be the leader of the local Freemasons, a mysterious band reportedly made up of "apostates from the faith, heretics, blasphemers devoted to the antichrist [and] about whom incredible tales were disseminated that they kill their enemies from a distance of several thousand versts through inexplicable means and such like fantasies."[3]

Such ideas about the Masons appear to have been rather common even before the reign of Catherine the Great, when the movement began to attract large numbers and spread throughout the country. The chief source for these views under Elizabeth was most likely the Orthodox clergy who saw Freemasonry as an expression of the harmful influence of increased contacts with the West. More precisely, Freemasonry was seen as a symptom of the anti-Russian outrages committed by the Germans who controlled the court under Empress Anne (1730–1740) (the so-called *Bironovshchina*), and as evidence of the continued dominance of the "German party" in the brief reign of Ivan VI (1740–1741). The leading clergymen greeted the coup of 1741 that put Elizabeth on the throne with a flood of praise for the new empress and an unending stream of vitriolic attacks against the deeds of the Germans Andrew Osterman, former head of the foreign office, and Count Burkhard Münnich, former head of the army, along with their treacherous Russian allies. These men were accused of robbing Russia's treasury, of forcing into exile all patriots and faithful servants of the state, of waging war against the Russian Orthodox church, and of giving free reign to the enemies of the true faith—namely, atheists, Mohammedans, and Armenians. Throughout the 1740s, the Russian clergy excoriated the "beastly, shameless atheists, apostates, schismatics, Armenians" and persons of "Epicurean and Masonic temper and mind," who had been allowed to rule Russia for too long and whose downfall Elizabeth's rise to power signaled.[4] It appears that to highly placed members of the church Freemasonry figured almost as part of a larger conspiracy directed against Russia by a band of Germans who, having weakened the country through these insidious imports, sought to control both state and society for their own gain.

The connection between Germany and Freemasonry appeared once more in the same decade in the interrogation of Count Nikolai Golovin. Upon returning to St. Petersburg in 1747 from Berlin where he had been in service to Frederick II, the young Golovin, suspected of carrying out a secret mission in the capital for the Prussian king, was questioned by the head of the secret police, A. I. Shuvalov, on a variety of subjects, including his affiliation with the Freemasons, his knowledge of the order's Russian membership, and the character of the society's laws and statutes. Although no disciplinary action was taken against Golovin as a result of his membership in the brotherhood—or against the Russian Freemasons he named—the circumstances surrounding the case suggest that the perception of the order as a suspicious foreign, and specifically German, presence was shared by part of the ruling elite along with the Russian clergy.[5]

It was about the same time that written attacks against the Masons began to appear.[6] Of these anonymous lampoons, which circulated among Russia's readers in hand-copied manuscripts, two stand out: "A Declaration Concerning Some Famous Deeds of the Accursed Masonic Mob" and "A Psalm Exposing the Freemasons."[7] The former, a lengthy diatribe in verse of over two hundred lines most likely composed by a Russian clergyman, depicts the Freemasons as Satan's lackeys and chronicles in fantastic detail the supposed diabolical ways of this evil sect.[8] It begins:

> In Russia Masons have appeared recently,
> And they're making demonic laws almost openly.
> Insidiously they try to hatch all sorts of tricks,
> In order to lure Christians to the antichrist.[9]

Having grabbed his reader's attention in this dramatic manner, the author—who freely admitted that much of what he was about to tell was perhaps already known to many—instructed those who wished to learn more of the Masons' rites to read on, promising to make known exactly how the Freemasons "secretly oblige the prince of darkness."

Next, the satire paints a terrifying picture of the Masonic initiation ritual. The candidate is first taken to a dark chamber where he is forced to endure numerous horrors. As dead bodies with chattering teeth mysteriously rise up from their graves, men run at the candidate from all directions, tearing his body with "pincers" and hacking at him with "swords and knives." He is then led to an underground pit to which the light of day has never penetrated. Surrounded by the ceaseless wail of impious, ghoulish voices and bathed in eerie candlelight intended to "increase the terror" of the setting, the aspirant, looking over a skeleton lying in an open grave, is handed a "glass full of evil," forced to disrobe, and then buried alive for three hours as a test of his bravery. Having withstood this trial, the candidate signs an oath with his own blood. Finally, each new brother has his portrait painted, after which it is removed to a special chamber.[10]

According to the author, the reward for withstanding this horrific initiation is the right to partake in all manner of dissolute and debased acts and a complete release from any moral constraint. The initiate has crossed over into a world turned upside down; he finds himself in a setting where evil has taken the place of good, wrong that of right, sin that of virtue.[11] Having apologized in advance for the "indecent" lines he is about to present, the author goes on to describe this perverted world:

> Here fine fellows and maidens sing as a choir,
> To a hidden room they retire to play French horns.
> They cavort there and dance with quick steps,
> And kiss wrapped in each other's arms.
> They celebrate by drinking various wines,
> Diverse dishes are set out on a well laid table.
> Eat what you will, who've been called to Masonry.
> . . . .
> Everyone runs off to special chambers,

And carefully puts on costumes.
Here Sodom and Gomorrah are present,
Here every sin and nonsense are villainously added.
Youths have made themselves over as maidens,
They've become real ladies and changed into dresses.
They dishonor themselves there with whomever they can catch;
Relatives don't avoid each other, fearing no sins—
Let it be one's sister or mother, just let us grab her.
Their wicked science leaves no one at peace.
She who doesn't wish to engage in carnal acts
Jumps aside to the wall, saying:
Now I've fled to a quiet refuge,
Neither the small nor great sin have I committed.
They'll tell her such hollow pretense is inappropriate—
Greetings, hour-long bride of our sect.
She's forced to say hello dear groom,
Kindly possess me now, I submit to your desires.
Through such deeds do the evil Freemasons
Construct with their own hands the hills of Babylon.[12]

Immoderate revelry, drunkenness, gluttony, fornication, sexual perver-
sion, incest, rape: these are the real deeds of the Freemasons, this is the ugly
reality that reigns inside their "temple of lawlessness" where they busily
erect a "throne to the antichrist."[13] As men of the utmost debauchery, slaves
to their "passions" and to the unquenchable thirst to experience whatever
their "flesh desires," the Masons know no boundaries and recognize no limits to their godless actions. Not content with debasing merely themselves,
the Masons actively seek to "entice all Orthodox Christians / And snare
them through cunning and deception for the devil."[14] The verses warn
readers against joining the order out of curiosity. Such a step, they are told,
could well produce tragic results since the brothers' portraits are intended
for one specific purpose:

There are said to be many examples
Of those who wished to quit this faith,
But none are on earth among the living.
Life and death, you see, reside in his portrait,
Which they'll simply shoot with a pistol,
And immediately, he'll languish and die.[15]

Such a pitiful fate, however, is preferable to the one slated for the rest of
the order. Soon all of their festive music, merrymaking, and fancy dress will
be gone and the entire body of Freemasonry along with them. They have

condemned themselves to an "eternal death" in hell from which there is no escape. Once these slaves of the devil have been committed to the flames, Hades will ring with the "clamor of [their] expiration" whilst they, "honorable people" and true Christians, rejoice in the triumph of their faith and curse the name of Freemasonry.[16]

Whereas this early doggerel most probably sprang from the milieu of the Russian clergy, the second squib, "A Psalm Exposing the Freemasons," appears more likely to have been the work of a Russian belletrist dating from the 1750s or early 1760s.[17] Unlike the prolix "Declaration," the "Psalm" is a model of brevity, conveying its disdain for Freemasonry in a few terse verses:

> Your Masonic laws
> Have proven to be full of lies,
>> And your secret is
>> The number six hundred sixty six.
> The grand ways of luxury
> Give rise in you to all vices.
>> Your assembly is there,
>> Where there's six hundred sixty six.
> Who in French is called a 'maçon,'
> Is in Russian a 'mason.'
>> There is no more ignoble name,
>> Even six hundred and sixty six.
> In whom there is real virtue,
> All people are witness thereof.
>> They'll weave a crown of praise,
>> Against six hundred and sixty six.
> And he who's ashamed of his actions,
> Hides everything from the world,
>> Until fate takes its vengeance,
>> Against six hundred sixty six.[18]

The major themes of the anti-Masonic "Declaration" are repeated here: the claim that the Masons' law is corrupt, false; the equating of the Masons' affairs with the work of Satan—in this instance represented by the number "six hundred sixty six," the numerical sign of the devil, chosen perhaps as a subtle reference to the Masons' fondness for numerology; the charge that Masons are responsible for creating all kinds of vice; the suggestion that the Masons operate in secret since evil and wickedness, unlike virtue and morality, must conceal themselves behind protective cover; and finally, the intimation of the order's pending doom. Also worthy of note in the "Psalm" is the direct reference to Freemasonry's Frenchness, a theme that while also present in the "Declaration" figures there much less prominently.[19] The fact that Freemasonry originated in Britain and that the lodges arrived in Russia via Englishmen, Scots, and Germans, as well as Frenchmen makes the emphasis on Freemasonry's Gallic nature somewhat surprising. While this may be a reflection of the particular importance of France as a source for Russian Masonry, it is also quite possibly an expression of Russian hostility toward France as the traditional ally of Russia's enemies (for example, the Ottoman empire, Poland, and Sweden) as well as a sign of an emerging francophobia then making itself known in response to the Russian elite's pronounced Gallomania. By equating Freemasonry with the empty mimicking of French fashion, critics perhaps sought to place the movement on par with the craze for all things French.[20]

The "Psalm" circulated throughout the country in hand-copied manuscripts as did other anti-Masonic satires of the period. Yet while most of these works, cut off from a broad audience, were largely destined for obscurity, a different future awaited the "Psalm" when it appeared in 1769 among a collection of "Fashionable Songs or Cures for Idleness" in N. G. Kurganov's *Universal Russian Grammar, or a General Manual of Letter Writing, Offering the Simplest Method for a Thorough Instruction in the Russian Language with Seven Additions of Various Educational and Useful-Yet-Amusing Things*.[21] The significance of its inclusion in this collection cannot be exaggerated given the remarkable history of Kurganov's book. An instructor of mathematics, navigation, and astronomy at the Naval Cadet Corps, Kurganov brought together a wide range of information to create a sort of encyclopedia or compendium of useful knowledge. The *Universal Russian Grammar* had sections devoted to Russian history, philosophy, and the natural sciences; it included information on grammar, science, art; it contained collections of jokes, proverbs, riddles, verses, reference material, and a dictionary. Kurganov's book proved to be extremely popular with Russia's readers, and following its initial publication in 1769 it was subsequently republished (under the title *A Manual of Letter-Writing* [*Pis'movnik*]) in 1777, 1788, 1790, 1793, and 1796—eventually reaching eleven editions by 1837. Kurganov's *Manual* could be found in the library of almost every Russian family. It was, to quote one authority on the history of the Russian book, the "gateway to learning" through which the growing body of readers passed.[22] Thus, it may not be an exaggeration to say that for Russians in the final decades of the eighteenth century, learning to read included imbibing anti-Masonic literature. The publication of the "Psalm" in Kurganov's compendium meant that not only were these derogatory images of the Freemasons spread to even the lowest levels of Russia's reading public—from one end of the empire to the other—but also, given the common habit of reading aloud in small groups, that these images reached persons outside the public as well.

. . .

Attacks such as these did not go unanswered, however. While their detractors worked to sully the image of Freemasonry, the brothers fought back by penning their own defenses, which also circulated in manuscripts among Russia's readers.[23] Sometime around the middle of the century, one unknown member of the order, apparently a young guards officer, kept a collection of popular romances, pastorals, and poems that contained a few rejoinders to the order's opponents. While repeating the standard refrain about the brothers as being men of moderation, honesty, and morality, these verses also exhibit a pronounced concern with responding to the Masons' critics.

> Stop cursing us, you
> Who've not joined the Masons.
> Try all you can, you'll never learn
> What the chosen one has found.
> > We pity everything about you,
> > This is why you curse us.
> > No, no, no, you won't understand us
> > As long as you aren't brothers yourselves.
> > No, no, no, you won't understand us,
> > Since you haven't understood us till this day.
> . . . .
> So, now you know our duties,
> And be ashamed of cursing us.
> Abide in your mores,
> And leave us to carry out
> > That which the law and honor dictate
> > And what we as brothers promised.
> > It's such, such, such a pity that you are not as we,
> > You'd spend your lives more agreeably.
> > It's such, such, such a pity that you all are not as we,
> > Then virtue would reign among you.[24]

These lines evinced once more the Masons' claim to social and moral superiority over those outside the order, in this instance antagonists depicted as ignorant of Freemasonry's true ways, lacking all virtue, and fated to persist in their debased, inferior mores.

Although the defense of the order initially fell to Masons with a modicum of literary talent, whose names have long been forgotten, the appearance of "A Psalm Exposing the Freemasons" seems to have provoked enough consternation and anger that none other than A. P. Sumarokov, the

Russian Racine, responded to the brotherhood's detractors.[25] In his short rebuttal, Sumarokov took direct aim at Masonry's outspoken opponents, defending the actions and character of the order, including its use of secrecy:

> Who criticizes the Freemasons
> For their secret rules,
> That they do not honor the laws,
> But adhere to their own idea of right—
> > If you were asked,
> > How loyal is the Freemason?
> > The law to which he is born,
> > Is the one he supports.
> . . . .
> People who are in power should
> Love the Christian faith so much
> That they are ready to withstand all attacks
> And to spill one's blood
> > In preserving the holy honor,
> > Not damaging it even a bit,
> > And revering the name of honor
> > Far above all vanities.
> To love people as one should,
> And to help the poor,
> And, as much as possible,
> To keep them from calamity.
> > I'll tell you in a word,
> > He's an honorable man,
> > But to learn their secret
> > Will forever be impossible for you.

These few examples suggest the degree to which Russia's Freemasons were aware of the extreme derision and condemnation directed their way by certain segments of Russian society. Despite the Masons' efforts to defend the order, by the beginning of Catherine's reign in 1762 Freemasonry had become embedded in the Russian mind as something nefarious and treacherous, an attitude succinctly expressed in a written denunciation from that year, which referred to its subject as "a suspicious person, a Mason."[26] The reactions Freemasonry provoked in Russia largely mirrored those present in the rest of Europe, where the order was linked with sorcery, licentiousness, irreligion, and subversion. Around the middle of the century, rumors became common in Holland, for instance, that the brothers played at "Devil's Cards" in their assemblies and that they possessed certain magical, occult powers making it possible for them to fly. A satire published in the

Boston Post-Boy (1750/1) implying that the Mason's trowel was actually used for anal tortures typifies the sort of sexual innuendo that was expressive of much early anti-Masonry. The fact that the lodges brought together men of various confessions outside the purview of the church and that the brothers made free use of religious forms and symbols for their own suspicious ends made them easy targets for the defenders of established religion. The foreignness of Freemasonry was yet another cause for alarm on the Continent. Part of the concern stemmed from Freemasonry's British origins, which made the lodges suspect in some circles as hotbeds of republicanism and political supervision in general. Still, as J. M. Roberts has observed, despite these initial fears, for the first half century of Freemasonry's life in Europe there was no "*general* ideological or political suspicion" of the order; only in the wake of the Bavarian Illuminati scare of the mid-1780s and the French Revolution did a fully elaborated mythology of Freemasonry arise. Rather, what one sees at this time are local, particular responses that, while sharing many features, did not add up to a unified, developed theory of anti-Masonry.[27] In this, too, Russia evinced a historical pattern similar to the rest of Europe.

Yet if one listens closely to the voices raised against Masonry in Russia, one detects in them a shrillness, a sense of outrage and fear greater than that heard in other parts of Europe. Not viewed merely as a threat to the church, Freemasonry was seen by many Russians as the work of the devil himself. There was something exaggerated or extreme in the Russian attitude that set it apart from other contemporary expressions of anti-Masonry. The reasons for this are difficult to discover because of the paucity of defamations and our ignorance of the circumstances in which they were written. Nonetheless, what evidence exists shows that at least in its first decades in Russia, Freemasonry was unable to shake off its alien origins. To most Russians, Freemasonry remained something uncomfortably foreign and, hence, dangerous. It turned out to be a perception of great longevity.

CHARLATANS AND DUPES

On 2 February 1786, the elite of St. Petersburg society gathered in the Hermitage Theater for the premier of a new play. This was no ordinary theater outing, however, for two simple reasons. First, the author of the evening's comedy, entitled *The Deceived*, was none other than Empress Catherine II, and second, the chief object of the play's satire included members of the audience itself—namely, the Freemasons. Set among the Radotovs—a representative family of Russian polite society—*The Deceived* opens to the image of a household turned upside down. Once a loving husband, caring father, and dutiful son, Radotov (Mr. Driveler) has undergone a radical transformation. He has become alienated from his own family and friends, preferring to pass his days alone in stony silence staring vacantly into space and, when deigning to speak, then only in a cryptic fashion, relying on odd words and expressions unintelligible to others.[28] A similar disorder has seized Radotov's daughter, Taisa. She has recently been seen gently kissing the leaves of plants, on whose tips, she claims, dwell thousands of invisible spirits.[29] Not even the servants are immune to this affliction. Praskov'ia, Madam Radotov's housemaid, has also acquired the habit of speaking in riddles unintelligible to her suitor, Tef. When he asks Praskov'ia where she gets "such refined words," she tells him that in her home "everyone has his own language" made up of all sorts of "queer words" that don't seem to make any sense.[30] These odd developments are too much for Radotov's mother to take, and she threatens to leave this house where "everything has gone topsy-turvy."

The source for this bizarre state of affairs turns out to be a suspicious necromancer and supposed wise man by the name of Protolk (Mr. Tattler), who has cleverly managed to worm his way into Radotov's home and has gained a powerful hold over his naive host through vague promises of personal enlightenment and communication with the higher realm of spirits. Radotov has begun meeting in secret with a mysterious band of men led by Protolk and his assistant (and accomplice), Bebin, who shares this "unintelligible language." Together they engage in all manner of esoteric rites and ceremonies, the ultimate aim of which—despite the artful façade of great solemnity and hushed seriousness—betrays a pitifully mundane and commonplace reality.[31] Setting aside their lofty talk, mock asceticism, and theatrical disdain for worldly affairs, the members of this motley band are actually driven by a hunger more material than spiritual in nature: just as Protolk and Bebin are drawn to Radotov out of a desire to dupe him of his wealth, Radotov seeks in them the knowledge of alchemy and the concomitant "priceless treasures" it promises.[32] Playing on Radotov's greed and his wish to appear enlightened, the deceivers not only rob him of his own money and papers, but also manage to make off with other people's funds, whose safekeeping had been entrusted to Radotov.

Radotov breaks down upon hearing of the theft. He realizes that he is financially ruined and has been publicly humiliated. Forced to explain himself to family and friends, Radotov confesses: "I was deceived by appearances; they both [that is, Protolk and Bebin] incessantly repeated how one is required to be virtuous according to their way of thinking."[33] He explains how like others he was initially attracted to these men out of simple curiosity, but then he was seized by the desire to appear better than those around him: "[M]y pride found satisfaction in distinguishing myself, in thinking differently from my family, my acquaintances." He was flattered to be in the company of men "full of knowledge, enlightenment," men who had "distinguished themselves from the ranks of ordinary people" through their superior virtue. Moreover, gullibility simply got the better of him; yet, he reasons, who wouldn't want to see and hear things that are generally considered impossible?[34]

Despite the multiple catastrophes resulting from Radotov's errors, *The Deceived* remains true to the classical formulations of its genre, and in the end order is miraculously restored. Protolk, Bebin, and the rest of the gang are arrested; the Radotovs' money is returned; and the remaining funds are turned over to a court to decide their disposal. In addition, the marital fate of the play's several young maidens—which Radotov had compromised by irresponsibly promising the hands of his daughter and his niece, Sophia, to these cheats—is settled to everyone's satisfaction as each is united with her true love. Finally, as the curtain falls, Britiagin, Radotov's brother-in-law and the play's primary voice of reason, steps to the edge of the stage and imparts to the audience the ultimate moral they ought to take home with them:

> Each age is judged by its descendants according to its sophistry . . . In general, praise is accorded only to those centuries that distinguish themselves through common sense, not through delirium . . . supervision is indisputably in the hands of the authorities . . . we must thank Providence that we live in such a time when gentle methods are used for correction.[35]

The delirium that *The Deceived* sought to correct might best be characterized as the prevalent susceptibility to appearances. The danger of appearances, that of seeming over being, runs throughout the play and serves to unite its various elements. *The Deceived* is filled with characters who—out of a longing to set themselves apart from their surroundings—simply adopt the signs of beauty, wisdom, and virtue in the mistaken belief that this will make them beautiful, wise, and virtuous. Like her uncle, Sophia is one of the comedy's best examples of this fallacious reasoning. She has arrayed herself in the latest (and most outlandish) French fashions and learned to mimic (to the point of caricature) the most exquisite Parisian manners in her rush to appear refined. Yet since these attributes do not emanate from any natural, internal source of elegance and grace but instead serve to hide this spiritual deficit, they remain forced and artificial, communicating the opposite of what Sophia intends. The appearance of beauty is no substitute for the real thing, and exaggerated attention to the outward expression of beauty is one of the surest signs of its absence.

This is the charge *The Deceived* levels against its primary target: Freemasonry. Another fashionable import not unlike Sophia's costumes, Freemasonry offers one the chance to appear enlightened and virtuous without necessarily being so. Their very public claims to the contrary, the Masons are no more morally advanced or spiritually self-disciplined than anyone else; in fact, quite the opposite is true. Radotov's pretensions to intellectual profundity and spiritual regeneration are empty gestures that only serve to highlight the depravity of Masonic activities. The lodge is a place in which the Masons can (and frequently do) abandon themselves to their basest passions and instincts (such as excessive drinking and eating) while hiding behind the banners of virtue and enlightenment. The Masons are not united by any shared love of morality, by a desire to strengthen the frayed bonds of humanity, or by a joint commitment to further the common good. Rather, what unites them are the vices of vanity, gluttony, folly, and greed. This makes Radotov's error much more harmful than Sophia's, since his search for social distinction led to such negative consequences. As *The Deceived* makes clear, by allowing himself to be duped, Radotov forsook his primary duties to family, friends, and society and risked doing irreparable damage to all. On stage such mistakes can be easily remedied; in life, however, they are infinitely more difficult to repair.

. . .

The Deceived is important for reconstructing the image of Freemasonry because it reflects a fundamental shift in this image, from that of the order as a demonic band of infidels to that of the order as a gang of crafty swindlers. Unlike earlier images of Freemasonry as the devil's attack on society's religious foundations, Freemasonry here appears as a purely secular phenomenon—an elaborate ruse used by clever rogues to rob the gullible rich. Although this view of Masonry cannot be attributed solely to Catherine, no other writer in Russia did as much to spread such a perception. In addition to *The Deceived*, Catherine penned two other anti-Masonic plays in the 1780s—*The Deceiver*, the first of the three, which premiered on 4 January 1786, and *The Siberian Shaman*, which opened on 24 September.[36] The debut of *The Deceiver* marked the appearance of Catherine's first play in over a dozen years and signaled the beginning of a fruitful period of playwriting for the tsarina. All three of Catherine's anti-Masonic plays are variations on a single theme. Each one tells the story of how a mysterious figure claiming supernatural powers manages to gain control over a well-to-do family (or at least over its male head), whom he seeks to trick for his own material advantage. In *The Deceiver*, Kalifalkzherston, a charlatan passing himself off as a great healer, alchemist, and necromancer, tries to cheat the Samblins out of their money and jewels through various hoaxes. In *The Siberian Shaman*, the Bobins have brought back to the capital from Irkutsk a Siberian priest, Amban-Lai, who uses their home to establish himself as a powerful oracle and seer. Of course, neither succeeds. Kalifalkzherston is caught with his accomplice, Madame Gribuzh, while absconding with their loot, and Amban-Lai is arrested after word gets out that he fooled a merchant widow into believing a bearded stranger was really the spirit of her deceased husband.

Following their debuts at the Hermitage Theater, Catherine's comedies quickly reached a much broader audience playing to the large crowds at the public theaters in St. Petersburg and Moscow throughout 1786 and 1787.[37] All three works were also printed separately and sold in local bookstores. First published in 1785, *The Deceiver* sold out completely at St. Petersburg's booksellers in a mere three days.[38] A second printing of *The Deceiver* was released only weeks later, at the beginning of the following year, along with two printings of *The Deceived* and a single edition of *The Siberian Shaman*, which also sold out in a matter of days. The empress quickly commissioned translations of the works, and by the end of 1786 German editions of the plays had been published as well.[39]

Although Catherine's plays were staged and published anonymously—perhaps in an attempt to direct attention away from the messenger and toward the message—the outpouring of public praise that followed their appearance left little doubt that her authorship was no secret. The weekly *Mirror of the World* for 9 February 1786 contained a glowing review of *The Deceiver* noting how "the public, having seen the above-mentioned several times at the theater, had established its merit with its incessant exclamations and uninterrupted applause." The reviewer lauded the play's publication as this would allow those living outside the capital the chance to benefit from its comforting "moral admonition" and to enjoy the undeniable "spirit of the *great Molière*" that permeated the entire work.[40]

The February and March editions of the *Growing Vine* included numerous short pieces in praise of *The Deceiver* and its creator. "A Letter to the Author of the Comedy *The Deceiver*" applauded the play's role in helping to destroy "those astonishing and magical edifices constructed out of grandiose words." The anonymous letter writer conceded that there would be individuals who would reject the images depicted in this work, asserting that the Masons gather only to do good. But he who maintained this, the writer argued, "attaches to this [assertion] vanity worthy of condemnation or, what is even worse but quite frequent, understands this solely as *his own good*." Finally, the letter's author, adopting an air of national pride, submitted that while these refined pretensions may well be at home among Europe's more polished nations, clearly they had no place in a simpler, more traditional Russia: "Such conceits are all right for a people as shrewd as the French. There everyone can be healed with magnets, communicate with spirits, and predict the future. But to us *simple children* the only thing intelligible should be that which displays plain common sense. Our fathers stuck to this rule, and I don't know about others, but I have no cause to be ashamed of my forebears."[41]

A poem dedicated to *The Deceiver*'s playwright honored him (that is, her) for bringing to light the true goal of these charlatans who accepted into their ranks "practically only the rich":

The curtain, concealing the deception of people's minds
Where fraud's temple is built with the shroud of loud words,
You fearlessly open for fellow-citizens' benefit.
Before the darkness you kindle an undiluted light:
May all clearly see by Your Comedy
That the depth of priests' knowledge, their very worth,
Rests solely on the common credulity of people,
And that in the peal of their word, but empty promises;
Their lone aspiration, through pockets to go rummaging,
Having shoved weak minds into a fog of lies most cunning.[42]

Catherine's play even inspired the dramatist N. F. Emin to write *The Sham Sage*, a five-act comedy patterned after *The Deceiver*, that also premiered in St. Petersburg late in 1785.[43] Once more the audience encountered a naive urban couple, Mr. and Mrs. Gullible *(Legkover i Legkovera)*, and a spurious wise man, Mr. Cunning *(Khitroum)*, who has charmed his way into their home by convincing them that he alone knows the secret path to wisdom, enlightenment, and the lofty sphere where the spirits dwell. Interested as the Gullibles are in making contact with the spiritual realm, their ultimate goals are firmly rooted in the mundane material world: while her spouse hopes the spirits will help him to acquire "wealth and ranks," Mrs. Gullible seeks their aid in becoming "a little bit younger and a little bit prettier." Together, the Gullibles hope that their esoteric pursuits under Mr. Cunning's tutelage will make them appear learned in the eyes of "enlightened society."[44] Mr. Cunning alone remains committed to reality's spiritual essence, albeit in his own particular way. "Money," he confesses in a moment of candor, "is the spirit of the world."[45] The play's references to "working [one's] stone" as the road to enlightenment, to "measuring human life with the compasses of truth," and to "smooth[ing] and level[ing] the heart's pores with a trowel of the deepest wisdom," as well as Mr. Cunning's self-proclaimed mastery of "the science of self-knowledge" and his use of "the compasses and hammer" in the practice of this science, all point to Freemasonry as *The Sham Sage*'s main object of satire.[46]

Whereas Emin had modeled his comedy after Catherine's *Deceiver*, the empress had modeled hers after Molière's *Tartuffe*. The importance of literary models in shaping these works, as well as the prevalence of stock dramatic features of the period (for example, servants whose simplicity and honesty contrast with their masters' vanity and corruption, intricate plot twists often revolving around mistaken or purposely disguised identities, telltale *noms parlant* like Mr. Driveler and Mr. Cunning), alert one to the danger of reading them as sociological sketches of Russian life.[47] Even so, this should not take away from the equally significant extraliterary context, for it is known from the author herself that both *The Deceiver* and *The De-*

ceived were inspired by and based upon a very real and very widely talked about contemporary—namely, Cagliostro.[48]

Arguably the most famous adventurer of the eighteenth century, Giuseppe Balsamo, better known as Count Alessandro di Cagliostro, arrived in the Russian capital in the spring of 1779 from Mitau, where he had won over the local elite through séances, alchemical experiments, displays of magic, and sheer force of personality.[49] As he had quite successfully done in numerous other European capitals, Cagliostro immediately set himself up as a healer, quickly attracting a large following. Soon all of St. Petersburg was speaking of the marvelous count, and the capital's most powerful and wealthiest residents turned to him to heal their afflictions. One of those with whom Cagliostro established friendly relations was I. P. Elagin, the courtier and leading figure in St. Petersburg's Masonic circles.[50] Elagin and other notables, such as Count A. S. Stroganov, one of the most prominent grandees in Catherine's reign and an active Russian Freemason, were attracted to Cagliostro in part by his supposed possession of the Philosopher's Stone and the Elixir of Life.[51] Through such contacts Cagliostro gained access to the city's Masonic lodges. As he had attempted before throughout Europe, Cagliostro—with the aid of his wife, Lorenza, who participated in his Egyptian rite under the name of the Queen of Sheba—hoped to convert the local Masons to his particular system. Curious yet skeptical of the mysterious Grand Copt (Cagliostro's *nom de maçonnerie*), St. Petersburg's Masons demanded proof of his supernatural powers and advanced wisdom before agreeing to come over to him. What happened next is not exactly clear, except that Cagliostro's attempt to conjure, with the help of a young assistant, proved to be a dismal failure, forcing the Grand Copt to flee the gathering amid accusations of chicanery.[52]

The debacle suffered before his Masonic brothers was soon followed by a widespread loss of public confidence, the specific reasons for which are rather murky. According to one version, Cagliostro had presented himself in the Russian capital as a Spanish nobleman in his country's service. As his reputation grew, news of the remarkable "Spaniard" reached the Spanish ambassador who, after some investigations into the "doctor," published announcements in the local press informing the citizens of Petersburg that there was no noble Spanish family by the name of Cagliostro, and there was no record of such a man ever having been employed in the Spanish service.[53] A second story presented a very different reason for the adventurer's downfall. Supposedly, after Cagliostro boasted of having healed the dying child of a wealthy couple, for which he received a sizable reward, word began to spread that the healthy child Cagliostro returned to the parents was in fact not their own offspring. Cries of fraud and deception rang out, eventually forcing Cagliostro to confess that he had returned a healthy child in place of the sick one who, he admitted, had died in his custody.[54] In either case, around the beginnning of March 1780, the Grand Copt and the Queen of Sheba, their reputations destroyed, abandoned the Russian capital for better prospects in Warsaw.[55]

. . .

Catherine's dramatic satires of Cagliostro and his devotees struck a sensitive chord within Russian society, setting off a public debate that lasted for several years. In 1786, just as her plays were being performed and published in Russia, unknown supporters—most likely Masons who perceived the criticisms of Cagliostro as a more general attack upon the lodges— jumped to the Italian's defense by publishing two separate translations of *A Treatise on Behalf of Count Cagliostro.* Cagliostro had recently penned the *Treatise* in an attempt to clear his name during the famous Affair of the Diamond Necklace involving himself, several courtiers, and Marie-Antoinette and centering around the mysterious disappearance of a massive necklace with a weight of almost three thousand carats.[56] The anti-Cagliostro forces immediately responded with the publication of a Russian version of *A Reply*

A séance of Cagliostro. The growing infatuation in the century's final decades with the mysterious and the occult provided plenty of opportunities for mountebanks like the Italian adventurer Giuseppe Balsamo (1743–1795), a.k.a. Count Alessandro di Cagliostro, who professed the power to, among other things, communicate with the spirit world. Cagliostro visited all the major capitals of Europe, including St. Petersburg in 1779 where he stayed in the home of I. P. Elagin and tried to win over the brothers to his Egyptian Rite. To skeptics like Catherine II, Cagliostro and his followers were proof of Freemasonry's inherent foolishness.

on Behalf of Countess de Valois La Motte to the Treatise of Count Cagliostro, originally published in Paris in 1786 by the countess's lawyer, Monsieur Doillot, at the height of the public outcry over the affair. The following year a Russian translation of Charlotte von der Recke's *News of Cagliostro's Infamous Stay in Mitau in the Year 1779 and of his Magical Experiments There* appeared. Both works portrayed Cagliostro as a cheat and a charlatan—not unlike the magnetizers Johann Georg Schrepfer and Johann Gassner—who had deceived people through a unique combination of "Scripture, magic (secret science), and Freemasonry." And in 1788 in Moscow, *Cagliostro Discovered in Warsaw, or An Authentic Description of His Chemical and Magical Operations Conducted in this Capital City in 1780* came out, promising to expose the "deceiver Cagliostro" who had for so long "been abusing the Public's trust."[57]

What is most peculiar about this debate is that it erupted some six years *after* Cagliostro's visit to Russia. How does one account for the timing of Catherine's plays that unleashed the debate? Why would she have bothered to mock the adventurer and his followers so many years after his departure?[58] It is widely argued that the impetus for the empress's plays was her visit to Moscow in the summer of 1785, where Count Ia. A. Brius, the local governor-general, may have brought to Catherine's attention the numerous activities of the Moscow Masons led by Nikolai Novikov.[59] Although possible, it is more probable that Catherine's attention was called to Masonry by events then transpiring outside Russia, namely, the Affair of the Diamond Necklace that landed Cagliostro and his wife in the Bastille in 1785. Russia's public avidly followed these events in the periodical press, which carried the latest news of the fate of Cagliostro and Lorenza, the Countess de la Motte, and Cardinal de Rohan.[60] Catherine shared in the general preoccupation with the scandal raging in Paris, and her letters from this period to the journalist Friedrich Melchior Grimm—then living in Paris—are filled with the empress's opinions of Cagliostro, whom she referred to as a "franc charlatan" (a pun on *franc-maçon*), and of the "odor of knavery" surrounding the affair.[61] Ever the enlightened instructress, Catherine was not content, however, to confine her opinions of this cause célèbre engrossing Europe's literate classes to her personal correspondence, but sought to make her views more widely known through her plays. With her dramatic works translated into other European languages and presented on domestic and foreign stages, Catherine could delight in showing that the Enlightenment did not flow simply from west to east, from the polished capitals of London and Paris to the backward outposts of Russia, but that it did in fact also flow from east to west, from the banks of the Neva to the banks of the Seine.[62]

As central as Cagliostro was to the debate on Freemasonry, his prominence and seeming omnipresence in the public imagination cannot be explained solely with reference to the spectacular exploits and undeniable audacity of this mysterious son of Palermo. For while Cagliostro's considerable (though fleeting) success was unique, his occupation was not. The eighteenth century witnessed a ceaseless parade of dubious doctors and healers, spiritualists and magnetizers, who artfully combined medicine and showmanship. They ranged from the Scotsman James Graham—whose Temple of Health, with its strange air pumps, electrical apparata, and Grand Celestial Bed reportedly capable of curing any illness, attracted large crowds in London in the 1780s—to the "Moon Doctor" of Berlin, whose preferred method of treatment involved exposing his patients to lunar rays.[63] Russia too hosted such physicians-cum-philosophers, many of whom clearly perceived enlightened Europe's eastern extremity as fertile ground for their cures and prescriptions.[64] To cite one example, on 27 November 1784 the *Moscow News* printed a lengthy advertisement for a Josef Maggi, proclaiming the foreigner's power to heal all sorts of ailments, from paralysis and convulsions to "women's sickness," through the use of a miraculous "electrical machine and its special attachment."[65] Around the same time in Moscow, the wife of a major-general Kovalinskii allowed people to visit during her regularly occurring trances and to take part in experiments in animal magnetism.[66] It was this general milieu that gave such resonance among Russian theater audiences to Aleksandr Ablesimov's comic opera *The Miller: A Wizard, a Cheat, and a Matchmaker,* which told the story of a common miller who also operated as a phony soothsayer, making it one of the most popular and oft-performed works of the century.[67] Sometimes these fakes became notorious enough, and their activities suspicious enough, to attract the attention of the highest authorities. Catherine's letters to local officials betray a concern with tracking down various charlatans operating under assumed names and titles. Her missive from 11 September 1787 to the governor-general of Moscow, for example, warned him to be on the lookout for a "Polish Count Shidlovskii," a "tramp and deceiver" known to engage in all sorts of "swindling and debaucheries."[68] Thus, Cagliostro served as a lightning rod for more general concerns about itinerant healers and mountebanks who, combining bits of scientific information with esoterica and then buttressing both with claims of extraordinary powers and fanciful personal histories, played on educated society's curiosity about the miraculous and the bizarre, in the murky intellectual realm where modern science and ancient mysticism converged.

Given their profound interest in secret knowledge, Russia's Freemasons frequently became involved with unusual figures about whom it was uncertain whether they were wise men or merely impressive charlatans. The unpublished memoirs of one Russian Mason relate how in 1786 he and two brothers met a certain Khariton Tsitseronol (or Tsitseronolia), a Greek "artist [and] maker of fake marble," who claimed to be a Rosicrucian and spoke convincingly to them of "secret sciences." Their curiosity piqued, they asked this Greek to be their teacher and pledged their loyalty to him as a superior. The brothers rented a room in a local house where they gathered every Sunday from midnight till dawn to listen to their teacher's disquisitions. After several meetings it became apparent, despite (or perhaps because of) Khariton's rather poor Russian, that their teacher knew little of what he claimed. Confronted with the brothers' suspicions, Khariton quailed and sought to defend himself. The brothers, realizing that they had been duped, abandoned their leader.[69] In a separate incident, A. A. Rzhevskii received a letter from Novikov in the winter of 1783, congratulating him on deciding to unite his St. Petersburg brothers with Novikov's group of Moscow Freemasons. In it Novikov writes confidently that their Masonic affairs will now run smoothly and properly since they "will finally be rid of all charlatanisms and will free themselves from the occasionally self-interested, occasionally ambitious, and occasionally erroneous aspects of certain brothers."[70] Several months later Prince N. N. Trubetskoi, Novikov's lodge brother, sent Rzhevskii new Masonic documents, informing him that, while not necessarily intended for use in their gatherings, they might be helpful so that "charlatans cannot cause any disturbances." Trubetskoi was specifically referring to José de Ribas, who claimed "knowledge of the internal order" but was actually nothing more than a "charlatan." By having these official documents in his possession, Rzhevskii would be able to test Ribas's true knowledge of the order. In another letter to Rzhevskii, Trubetskoi wrote that it was crucial for Russian Masons to take control of their lodges and to stop groveling before foreigners. The time had long since come, he urged, for them to stop "running after every little tramp who, being a nobody in the order, poses here as someone great" and to adopt a more critical stance toward all these "Ribases, Rosenbergs, Frezes, and their kind."[71]

One major problem all Freemasons faced was deciding who exactly the charlatans were. In sharp contrast to Trubetskoi, for example, Baron de Corberon, a devoted Mason and the French chargé d'affaires in Petersburg, described Monsieur Ribas as "the most honorable young man one could know,"[72] and while Novikov and Trubetskoi could easily judge brothers they disliked as frauds, they themselves were not immune from similar accusations. The memoirist A. T. Bolotov recalled resisting Novikov's attempt to lure him into the order. He recorded his thoughts upon leaving a meeting with Novikov thus: "No, no sir! You haven't descended upon such a fool and little simpleton who would simply let himself be blinded by your idle chatter and tales and would stretch out his neck so that you could put upon it a bridle in order to ride him like a horse and forcibly make him do everything that you please."[73] The same Freze whom Trubetskoi had labeled a "little tramp," Baron Heinrich Jacob Schröder referred to in his diary as "a rather good and honest man," and J. G. Schwarz, a trusted Masonic comrade of Novikov and Trubetskoi, was deemed by J. P. Wegelin, the orator of

Moscow's Three Banners lodge, to be "an expert deceiver."[74] Schwarz, naturally, believed he was one of the few honest foreigners in Russia. He described his feelings in an autobiographical sketch: "I became indignant seeing how unworthy, self-seeking foreigners deceive many noble fathers and mothers who ardently wish their children well, but do not have enough education to know how to go about matters." It was this regrettable state of affairs that supposedly prompted Schwarz to establish his own educational institution in Moscow. Fortunately for Schwarz, not everyone shared Wegelin's assessment of his character, and he was able to attract several patrons for his project. Among them was P. A. Tatishchev, a wealthy Russian and brother Mason who had himself, according to Schwarz, "been duped more than once by the throng of flatterers that always crowd around the rich."[75] If one believes Wegelin's version of their relationship, the worst flatterer of them all was the pseudoscholar Schwarz, whom he accused of defrauding Tatishchev of thousands of rubles.

. . .

*F*eared by segments of the public and by individuals in the lower social strata, ridiculed on the country's stages, mocked in the day's journals, pamphlets, and books,[76] and riven with deep divisions in their ranks, Russia's Masons responded by publishing their own interpretations of the order's true spirit and of its adherents' true character. On 22 August 1785, the *St. Petersburg News* carried an advertisement for one such interpretation, *Fraternal Admonitions to Some Brother Fr[ee] Ma[sons]*, printed twice in Moscow the previous year.[77] The book's author, S. S. Eli (or Ely), who published under the Masonic name of Brother Seddag, was a converted Polish Jew who lived in Russia from 1776, where he served in the College of Medicine and was a member of the Free Economic Society and of several Masonic lodges. Despite the title's implied readership, *Fraternal Admonitions* was directed at the general public and was not written solely for the edification of the movement's members. Eli sought to address directly the misconceptions and errors about Freemasonry within Russian society by stressing again and again the inherently benevolent goal of Masonic activity. He refuted all the charges against the order: that it served only "self-interest," its members simply seeking ways to make gold or acquire wealth; that it fed "blind ambition" by offering men a new sphere in which to chase after "different honors"; or that it harbored some "political secret."[78] Rather, Eli argued, Freemasonry was the proper place, indeed the best place, for all who wanted to "learn how truly to think, to discern, to choose, to want and to repudiate, to work and to act." Within this "school" there had gathered a "society of men founded for the sake of virtue, striving to learn irrefutably pure truths and to spread requisite and beneficial knowledge for the general good."[79]

The same year Eli's *Fraternal Admonitions* appeared, another important work was also published in Moscow—the *Free-Masonic Magazine,* a collection of speeches, discussions, songs, and choruses largely selected from local Russian lodges. According to the work's preface, the order published books such as *The Freemasons' Pocket Book; Fraternal Admonitions;* and *An Apology of the Order of Freemasons* so that "members of the Order as well as outsiders might be able to gather at least some basic ideas about true Freemasonry and put out of their minds highly erroneous conclusions and prejudices that false brothers and haters of the Order have for a long time been sowing into gullible and insufficiently steadfast hearts so as to harm the Order."[80]

The "highly erroneous conclusions and prejudices" that the *Free-Masonic Magazine*'s numerous speeches and discussions set out to refute were those discerned so far in various lampoons: that is, that Freemasonry was nothing more than childish games, that it was devoted to the satisfaction of vanity, and that it did nothing to better its members, their families, or society, but instead distracted their attentions and diverted their energies from truly useful activities. The *Free-Masonic Magazine* fully agreed that this was true of many "supposed" Freemasons. In "A Discourse on Authenticity," initially delivered in the Moscow lodge Radiant Triangle, the orator admitted that many attending Masonic assemblies were Masons only in name. These brothers visited the lodges "merely out of a sense of propriety," and they were present only in the most superficial sense: "Their bodies alone sit in this sanctuary, but their hearts are far away." The single reason they joined was to acquire the "social advantages and benefits" that came with membership in Freemasonry. The orator expressed complete confidence, however, that all false brothers would be duly exposed, assuring his listeners that "we have been attached to the true Order, giving us the means to distinguish the Brother from the charlatan and from the deceiver."[81]

But if false brothers sought to enhance their social status through lodge membership, what did real brothers seek? And how did this make them any different from and any better than Russians who avoided the lodges altogether? Answers to these questions can be found in a "dissertation" delivered to the brothers of the Holy Moses lodge. Devotion and obedience to the order, the brothers were told, "will awake in us true good behavior, characteristic of the true Freemason, and will make all our words and movements superior to those of the profane, presenting us as an example of social comportment."[82] They constituted a "small number of the wise," the "few true human beings," a fact demonstrated by the guiding role of "prudence" and "moderation" in their lives that distinguished them from the common man. Captive to the vices of pride, cupidity, and idleness, the common man wasted his days amid raucous gatherings where he surrendered himself to hedonism and drunkenness. In the end he became indistinguishable from swine. Having turned back the critics' charges against them, Freemasonry's defenders sought to depict themselves as society's most useful and benevolent members. Their energies were aimed first at fulfilling their "civic and domestic duties"; next, true Freemasons eschewed intemperate company in favor of "peaceful and reasonable conversation with like-minded friends bound together by brotherly love and pleasing comportment" such as could be found in the lodge.[83]

The link between the lodges and civic duty was most pronounced in another discussion from the Radiant Triangle. In perhaps the most arresting image of Freemasonry in the *Free-Masonic Magazine,* the lodge was equated with the gathering sites of the classical world and held up as the only enclave where civic virtue might still be cultivated and kept alive in an inhospitable, corrupt environment:

> Is the voice of the fatherland, having resounded with such strength in the assemblies of the Greeks and Romans, having thundered in the ears of the dying lovers of the fatherland, and having aroused in them a gratifying smile with their final breath, is this powerful voice already losing its force among us, or have we been deprived of their tender senses and become incapable of hearing it? We have no public places where citizens might gather to consult on the good of the fatherland; we have no Demostheneses, no Ciceros who might rouse our attention to this voice; we have no public images of men who have served and sacrificed their lives for the mother having borne all of us Russians that would declare to those gazing upon them: *die for the fatherland.* Will we now bar this voice entrance to its final refuge, to this sanctuary consecrated to truth and wisdom? Will it no longer fire their hearts with a burning love for their Sovereign, and will it no longer serve to spur us, in place of the trumpets' call to battle, to heroic deeds, deeds which if not immortal, are at least noble?[84]

Here is the lodge as ecclesia and comitia, and its members as citizens selflessly devoted to the commonweal, as loyal sons of the *patria,* and as true subjects of the ruler. Framing their activities in the language of civic humanism, these brothers of the Radiant Triangle sought to fashion an image of the lodge as a site for the gathering and strengthening of the nation's citizens. Here they could claim participation in the *vita activa* through their dedication to furthering the common good and their devotion to the native land.[85] Preoccupation with the self, with overcoming vice and promoting virtue, was a requisite first step in this program: by improving themselves, the Masons could in turn improve society. This activity had nothing in common with a "childish plaything";[86] nor did it have anything to do with a selfish, egocentric desire for wealth, power, or status. True Freemasonry's activities were directed outward, not toward one's own personal wants and needs, but toward those of society. That outsiders could not understand this essential fact testified to their own baseness and corruption.

What, then, is to be made of all these images, of this collection of dissemblers, alchemists, necromancers, dupes, loyal subjects, and good citizens? What is the common thread that runs through them all and ties them together? At bottom, what these images reflect is a shared concern over the question of Freemasonry's utility, its social usefulness. Reflecting the age's endorsement of Horace's companions *utile et dulce* (use and pleasure), their critics saw the Masons engaging merely in "amusement" *(uveselenie)* and not "utility" *(pol'za)*, and the forsaking of the latter in favor of the former made them worthy of condemnation. Mr. Rightthought *(Pravomysl')*, Mr. Gullible's wise and virtuous brother in *The Sham Sage*, expressed the view clearly when he instructed Mr. Cunning henceforth "to use himself for the benefit of society," which did not mean "try[ing] to penetrate the impenetrable," but "[s]erving the fatherland, helping one's neighbor."[87] In fact, this is what constituted true enlightenment, not attempting to "speak obscurely and unintelligibly . . . [and] to understand one's heart as a stone." The truly enlightened man was a "real citizen, son, spouse, and father who looks after the profit of others as well as his own." He did not waste his time conversing with "spirits" but was fully devoted to his "[o]ffice, economy, and the upbringing of children."[88] Unfortunately, the "knowledge" of swindlers such as Mr. Cunning had no "application and benefit" for society, but "corrupt[ed] morals and manners," precipitated the squandering "of one's property in a dissolute way," and turned "citizens into negligent, false wise men."[89] The Masons, on the other hand, claimed just the opposite, and through their books sought to influence public opinion by stressing the chief role of social utility in Masonic affairs.

That Russia's Freemasons needed to justify their gatherings in this manner was not unique. The lodges were not the only societies that were viewed disapprovingly at the time and that provoked disquiet. Indeed, the appearance of so many clubs, circles, and other assemblies in the second half of the century appears to have unleashed a good deal of worry over the benefit of such bodies and where such a development might eventually lead. When Novikov's journal the *Painter* announced the formation of the Society for the Printing of Books in February 1773, it stressed that the society was to be "the most useful institution of private persons of our century" and expressed the sincere desire that "information about such Society will perhaps prompt others to establish another, much more useful than our clubs, assemblies, and such like gatherings."[90] For individuals interested in establishing their own beneficial society to help the poor, yet fearful of being taken advantage of by "impudent spongers," the *St. Petersburg News* printed in April 1780 the guidelines used by a society in Germany that managed to extend aid only to "useful members of society" and not mistakenly to any "deceivers."[91] In his famous "Questions" to Catherine II—published in 1783 in the *Companion of Lovers of the Russian Word* (a journal dedicated to "amusement as well as true utility")—the writer and dramatist D. I. Fonvizin bemoaned the sorry state of Russian society, complaining that honest people were no more respected than idlers and that no stigma was attached to those who did nothing to serve society. Particularly disturbed by what he perceived to be the degraded condition of the nobility, Fonvizin asked the empress why in both capitals associations among nobles had become extinct, to which she answered: "From too many proliferating clubs."[92] With her succinct response, Catherine conveyed the pervasive concern about the questionable social impact of the growing number of sodalities. On 20 May 1783, the same journal published G. R. Derzhavin's "Felitsa," the poem that gained him the favorable attention of the tsarina and helped to launch his career in state service. In his satirical ode, Derzhavin praised the wise Felitsa (who represented the empress) for avoiding all masquerades, clubs, spiritualists' gatherings, and Masonic lodges, which were presented as the domain of lazy, corrupt, and hedonistic courtiers. In contrast to such indolent types, Felitsa was depicted as a figure of simple-yet-direct virtue, a model of rectitude and diligence free from all vanity and the dangerous grip of the passions.[93]

The border separating "amusement" and "utility" was fluid and ill-defined. Where it was located depended to a large degree upon where one stood. One person's useful activity was another's waste of time. Whether the growing collection of societies, clubs, assemblies, and lodges fed only Russians' appetite for entertainment was a matter of significant discussion. Amusement was good and not to be frowned upon, but when unaccompanied by any social benefit, amusement degenerated into something indefensible, worthless, and even harmful. And for Catherine, as for many other educated Russians, if the Freemason was sincerely motivated to do good for society, as he claimed, then "what need does he have of solemn oaths, of eccentricities, [and] of absurd and queer attire?"[94]

MARTINISTS

The pervasive notion that assemblies of all sorts were generally prone to mindless prattle and pointless activities served as the basis for Catherine's satirical "Daily Notes of the Society of Know-Nothings" published in 1783 in the *Companion of Lovers of the Russian Word*.[95] Attempting to lend an air of worldly sophistication to their assembly, the members called themselves the "Ignoranti Bambinelli," unaware that their chosen title meant "Ignorant Children." The society is shown, through the fictitious record of its meetings, to be nothing more than an exercise in absurdity as its members conduct meaningless discussions about insignificant and frequently incomprehensible topics. Proposals are offered, debated, sent back and forth between the society's two chambers, and then, following this flurry of activity, usually dropped or forgotten.

One pressing matter to which the Ignorant Children turned their attention was trying to uncover the meaning of the word *Pifik*. The representatives of neither chamber could agree on the proper meaning of the mysterious term. Some suggested the word referred to the natural sciences; others claimed it had been fabricated by envious persons for their own gain; still others interpreted the word as a derivative of spades *(piki)*, a foreign group known to have frequently engaged in combat with the diamonds *(bubni)*, the hearts *(chervi)*, and the clubs *(zhludi)*.[96] Not surprisingly, the Ignorant Children were unable to arrive at any consensus, and the matter remained unresolved. Anyone still interested in learning the answer, however, only had to consult the *Dictionary of the Russian Academy* and read the entry for "Martyshka":

> 1. Circopithecus. Pifik, a monkey with a long tail, with bare hind calluses and with cheek pouches. *A green martyshka.*[97]

While this solves one question, it raises two others: first, why did Catherine choose this odd monkey as a topic of discussion for her make-believe society; and second, what do monkeys have to do with Freemasonry? At first glance the empress's decision appears purely random, but there is more to her selection than chance. This is not the only place where "martyshki" appear in Catherine's writings, and by following the trail of monkeys, so to speak, one might discover how it leads to the Freemasons.

In the July 1786 issue of the *Growing Vine*, Catherine anonymously published "A Household Note on the Contagion of the New-Fangled Heresy and on Remedies for Recovering from It." The short piece begins:

> It is well known that new heretics have appeared in Russia, Martyny or Martyshki, I don't remember what they're called. But I know that their heresy is catching and a great number of the faithful have been infected by it. Even I, calmly roaming around the big wide world, would've gotten caught up in it, like a fish in a net, if I hadn't been saved by my doctor who, refusing to allow the illness to develop unchecked, with difficulty knocked this nonsense out of me . . . through various curative remedies.[98]

The author traced the cause of this illness back to the same source that had recently infected many others—the book "on truths and errors" [*sic*] over which he had been "breaking his head without mercy" in a vain attempt to understand its incoherent message. His blood inflamed, his eyes glazed over in a blank stare, foam forming at the corners of his mouth, the author was saved at the last minute by his doctor who, seeing the dangerous

book in his hands, threw it out the nearest window and immediately prescribed for his patient a five-step cure. For sixteen days he received a "cool-
ing agent" that trained him "to be with [his] wife, children, and acquaintances and occasionally to engage in innocent amusements," all of which he
had earlier neglected in favor of "these strange *Martinists* or *Martyshki*."
Next, he was given a "sickening agent" as a vaccine comprising, in part, "4
and a half ounces of Platonic rubbish, 6 and 1/4 ounces of Appollonian ravings, 10 ounces of the Jewish cabala's foolishness, and two full doses of the
wild madness of the renowned cobbler and philosopher Jacob Boehme."[99]
The treatment's third stage, the "instructive agent," required the patient to
read carefully both *The Deceiver* and *The Deceived,* the effect of which made
him laugh at his former follies, realizing how he and those like him were
just like Samblin and Radotov. For the penultimate stage, the "strengthening agent," he memorized the Ten Commandments since they contained
"the very best philosophy, teaching us to be worthy of and useful to society
and, consequently, pleasing to our Creator." The doctor himself conducted
the final, decisive operation, the "smoking treatment." First he took from
the patient's library every work by Swedenborg, next he removed the copy
of *On Errors and Truth* as well as other "heretical books," finally, he burned
them all. The smoke from the fire literally "fumigated" the recuperating author just as "fashionable women suffering from hysteria are fumigated with
burning feathers." Now healed and safely inoculated against future attacks,
the author advised everyone still infected with the disease to follow the described treatment and to consult any doctor easily found "in every city and
village in the territory of *common sense*."[100]

. . .

Not monkeys after all, this strange band of heretics owed their name to
Louis Claude de Saint-Martin, the French theosophist, Freemason, and author of *On Errors and Truth, or Man Restored to the Universal Principle of Science,* which Catherine blamed as the chief source of the mental disorder
supposedly sweeping the country. A combination of Christian mysticism,
the writings of Jacob Boehme, cabalistic numerology, and Masonic ideals,
On Errors and Truth (1775) made its way to Russia before 1780 and met with
immediate success.[101] A lengthy advertisement in the *Moscow News* in 1785
heralded the publication of a Russian translation, praising the book as
"epoch-making in the philosophical history of the most recent times" and
highly recommending it to all "our gentle compatriots."[102] The Russian version proved to be as popular as the French original, winning for itself ever
more avid readers and selling at a brisk pace.[103] Its popularity is surprising
given the book's impressive length and dense, cryptic language, which
must have discouraged many casual readers and tested the patience and
stamina of even the most enthusiastic. Even Elagin, a very supportive critic

who judged *On Errors and Truth* to be "the most important wisdom," required extensive exegetical help from brother Masons in his effort to unlock its secrets. Those who did not, could not, or simply would not understand *On Errors and Truth*, Elagin noted, dismissed it as "absurd nonsense and folly" worthy of laughter, or as an expression of a "secretive Jesuit system" dangerous for the state and its leaders, or as the work of the "illuminati."[104] One skeptic, the empress's court physician H. M. Weikard, wrote in a letter to Doctor Zimmermann in February 1786 that Prince N. V. Repnin and others had become enthusiasts of *On Errors and Truth:* "They say that only the profane do not understand it; but neither do the good gentlemen. They do not realize that it is foolishness."[105]

If there was little agreement over the meaning of *On Errors and Truth,* there was perhaps even less over who the Martinists were. Although the label eventually came to be almost exclusively associated with the circle of Moscow Masons centered around Novikov, throughout the 1770s and early 1780s it did not refer to any single group. Rather, "Martinist" was used indiscriminately as a general term of derision: to call someone a Martinist was an easy way to cast aspersion on him.[106] Equally vague was the relationship between the Martinists and the Freemasons. Although the two were not synonymous, by the 1780s they had become linked in the public imagination, and the image of the Martinist figured prominently in the plays and articles decrying Freemasonry as the playground for cheats and swindlers. In March 1786, for example, the *Growing Vine* printed an article on the "life of the first founder of the Martinist sect," a certain "Postell" from the sixteenth century who, although a truly gifted man, turned into a "deceiver" and an "extravagant lover" from reading too many "rabbis" and staring at the planets for too long.[107] And in the opening scene of *The Deceiver,* the housemaid Mar'ia tells a visitor that her master secretly meets with some strange men: "Mif . . . mish . . . mid . . . myt' . . . miar . . . mart . . . marty . . . I almost said with martyshki." "Ha, ha, ha!" he laughed, "What sort of martyshki? Imitative monkeys, even they are poseurs. What sort of people? Are they affected?"[108]

Prince P. D. Tsitsianov attended one of the initial St. Petersburg performances of *The Deceiver,* which he described in a letter to his friend V. N. Zinov'ev: "It ridicules those who believe in charlatans and especially believed in Cagliostro; it deals with a certain sect of 'Martinists' who converse with spirits and are quite famous here."[109] A few weeks later in response to Zinov'ev's queries, Tsitsianov wrote at some length of this mysterious and much talked-about "sect":

> The greater part of people here have been damaged by one sect called the "Martinists' Sect." The mask hiding all this hypocrisy has been stripped away. They read sacred books, the most ancient, and appear to be attached to God

(which, however, I doubt, seeing that their way of life, their behavior, and their affairs do not concur with true, God-revering wisdom, with His goodness, and His paternal care for us), then they appear estranged from this world and to be engaged with spirits—even more, in a word, than Platonists. Tout est spirituel. Books, that is "Des erreurs et de la vérité" and "Les merveilles du ciel et de l'enfer," they consider as their dear leaders, the first of which has even been translated into Russian, but I could not understand it not being enlightened, that is, a Mason . . . It (that is the first book) must be known to you since, I think, even there it is in vogue . . . What after all are these Martinists doing here deceiving fools from behind this mask of piety? They too lead debauched lives as do others who outwardly appear less pious; they too divorce their wives; they too keep a lover along with their wives; they too oppress the poor; they too rob the weak and may also be considered the scourge of the human race. And so I must conclude: either this sect is a work of knavery and a wicked mind covering itself in piety for its own gain, or it is good intentions that are being used for evil. That is everything I wanted to say about the Martinists . . . maybe I am mistaken; however, I accept good principles and a clean conscience as the object of piety.[110]

Tsitsianov's portrait is especially valuable for what it says about contemporary perceptions of the so-called Martinists. According to Tsitsianov, hypocrisy was their foremost feature; it tainted everything they did and said and everything they claimed to be. He maintained that the desire for social distinction underlay the Martinists' purported communion with the spirit realm, their alienation from the mundane, quotidian world, their immersion in holy, esoteric texts, and their pretensions to an exceptional piety—all of which he saw as nothing more than empty gestures, fashionable poses intended to impress those around them. But if anyone has been intimidated or deceived, Tsitsianov noted, then it had only been the members of this "sect" who, too naive and impressionable to know any better, had been duped into believing this entire charade by the likes of Cagliostro and his ilk.

The image of the Martinist, together with that of the Freemason, underwent an important transformation with the outbreak of the French Revolution in the summer of 1789. Literate Russians followed with avid interest the upheaval raging in France through the lurid reports in the *St. Petersburg News* and *Moscow News*. Publishers of Russia's newspapers could barely keep up with the swelling demand for information: the *St. Petersburg News* rose from a circulation of 970 in 1787 to 2,000 in 1790, and the *Moscow News* climbed to 4,000. A flood of revolutionary publications and tracts quickly made their way to Russia as well, as the country's booksellers busily sought to satisfy the unquenchable thirst for news. Everywhere—at court, in the homes of the nobility, in salons, and in polite circles—Russians talked of nothing but revolution.[111] Reaction to the revolution was not uniform. On one hand, it met with great enthusiasm and excitement, less perhaps for its social and political significance (which was not necessarily clear), and more for its novelty, for the sense of fun and fashion that seemed to be, at least from the safe distance of Moscow and St. Petersburg, a part of the whole experience. The revolution's sartorial styles, its specific language, its songs, and its expressions of camaraderie soon made their way to Russia, where they were hungrily taken up by the Russian public as the latest wave of French fashion. Publications like the monthly *Journal of New English, French, and German Fashions,* with its stories on the most recent French styles and on the country's "newest mode of life," fed the appetite for revolutionary dress. Local merchants advertised the new clothing in the two capitals' major newspapers.[112] The fascination with revolutionary material culture proved more than a brief infatuation, and as late as October 1792 women attended Moscow's Noble Assembly wearing "Red Jacobins' caps."[113]

Significant as this response to the revolution was, it was soon followed by, and existed uncomfortably alongside, yet another reaction: fear. The same books and newspapers that advertised the revolution's fashionableness (the first instance of radical chic?) simultaneously played to their readers' voyeuristic desire to witness the brutal killing and destruction then tearing apart the country. The Russian public feasted upon a steady diet of gruesome stories describing the "ruinous consequences of the French anarchy." They read published letters from the French provinces that detailed how the castles of countless nobles were being plundered and set afire and how the countryside flowed with blood from ceaseless murdering. The *St. Petersburg News* summed up the terrible situation existing in France in the spring of 1790: "[O]ne half of his majesty's subjects is exterminating the other with the most brutal tortures and assassinations. In the provinces the mob rages."[114]

Not long after the outbreak of the violence, blame for the unrest was laid on the country's numerous associations, political clubs, and especially Masonic lodges. That the cause of the upheaval was increasingly attributed to the Freemasons was not particularly surprising nor was it difficult to explain, for the lodges had long been viewed with suspicion throughout Europe. The fact that the lodges brought together men of differing confessions and that they used religious symbolism for their own mysterious ends had traditionally been perceived as a direct challenge to the established authority of the church, whose own status was deemed crucial to the maintenance of the social order. In France the connection between Freemasonry and religious subversion became more explicit as French clerics began to argue that the order was actually a Protestant movement aimed at destroying Catholicism.[115] The idea that Freemasonry was part of some larger Protestant plot owed much to the widespread conviction in many segments of European society that the lodges had arisen during the English revolutionary experience of the seventeenth century and that their true founder had been none other than Oliver Cromwell. Thanks to this fanciful myth, Freemasonry became associated not simply with Protestantism but, perhaps more significantly, with revolution, republicanism, and constitutional government. Indeed, the notion of the order as a revolutionary force allied with the spirit of Cromwell began decades before the French Revolution.[116] To these older threads of anti-Masonry the revolution's critics added more recent ones to weave a complex tapestry of grand conspiracy that combined Freemasons, Jacobins, Martinists, Illuminati, the philosophes, and Rosicrucians, and which found its most elaborate composition in the abbé de Barruel's *Memoirs, Illustrating the History of Jacobinism* (1797), what J. M. Roberts has called "the bible of the secret society mythology and the indispensable foundation of future anti-masonic writing."[117]

Rumors about the Freemasons' responsibility for the disorder in France were already spreading throughout Europe before the end of 1789. These rumors found especially fertile soil in Germany and the Habsburg lands where, since the Illuminati panic of the mid-1780s, the lodges had been suspected of harboring political subversives.[118] In August 1790, a Viennese official reported to the emperor that he could prove that "the wheel of the present errors and revolutions in Europe was being propelled by the brotherhood of the Freemasons." In October of that year Hamburg's *Political Journal* wrote of a "Club de Propagande" that supposedly met twice weekly "in the manner of the Freemasons" in order "to spread the revolution throughout all of Europe."[119] This report would have come as no surprise to tsarist authorities, who as early as the spring of 1790 had begun worriedly discussing the establishment of a "special club" in Paris, created to carry the revolution to the rest of Europe, and by the next year Russia's newspapers had singled out the "Jacobin club" as a major source of the present evil.[120] In the same year in Moscow, Prince N. N. Trubetskoi read to his dismay in the *Mercury of France* the review of a new French journal, *Bouche de fer* (Iron Mouth), whose editors, "posing as Masons," actually sought to take credit for the revolution.[121] Some thirty years later, A. M. Turgenev recalled that during this period it was rumored in society that the Jacobins and the Freemasons had even united to try to poison the Russian empress herself. Neither limited to the capital nor furtively spoken of in hushed tones, this rumor supposedly reached the most remote corners of the country and was openly discussed everywhere—"in conversations, at markets, in a word, in all gatherings."[122]

Important as the revolution was for transforming the public perception of the Freemason and the Martinist, it appears that even before 1789 these figures could excite overtly political suspicions. During a visit to St. Petersburg in March 1786, Friedrich Tieman, an influential Mason and friend of

Johann Casper Lavater and Saint-Martin, wrote in a letter to his fellow Freemason Jean-Baptiste Willermoz that the Martinist sect in Russia, whose ranks he estimated at more than three thousand, was "imputed to follow the principles of Cromwell and to be planning a revolution."[123] Intriguing though Tieman's description is, its validity (especially given his wildly exaggerated figure for the number of Russian Martinists) is difficult to assess, and it appears safe to assume that, until gripped by the fear of revolution, most Russians understood the Martinists more as contemptible yet harmless pseudo-spiritualists than as dangerous subversives. The scandal over A. N. Radishchev's *A Journey from St. Petersburg to Moscow*, published in May 1790, neatly conveys the shift from the former to the latter. Even though Radishchev explicitly mocked the Martinists as austere, unfeeling ascetics who would rather bury themselves in "Hebrew or Arabic letters, in ciphers, or Egyptian hieroglyphs" than "spend the night with a pretty girl . . . intoxicated with passion," Catherine, upon reading the *Journey*, labeled its author a Martinist, though in this newer sense of the word:

> The purpose of this book is clear on every page: its author, infected and full of the French madness, is trying in every possible way to break down respect for authority and for the authorities, to stir up in the people indignation against their superiors and against the government.
> He is probably a Martinist or something similar.[124]

Others apparently detected the harmful spirit of Martinism in Radishchev's book as well. In October 1790, I. V. Lopukhin wrote to A. M. Kutuzov—to whom Radishchev had dedicated his *Journey* and who was then attending to Masonic affairs in Berlin—that in Moscow "it is said that he [Radishchev] is a Martinist." This attribution surprised both Lopukhin and Kutuzov given their knowledge of the distance that separated them— as sympathetic readers of Saint-Martin—from Radishchev, who was openly critical of him.[125] The reason for the confusion, according to Lopukhin and his fellow Masons, was that the public was not able, or not willing, to see any distinction between the various groups. Moscow society, Lopukhin noted, lumps all these groups together—"the Martinists, the Illuminati, and the Masons for the sake of abusing them." Lopukhin claimed not to know anything about any Illuminati in Russia or whether a society of Martinists even existed. He was sure, however, that whether or not Radishchev was a Mason, his book had nothing in common with "true Masonry."[126]

Kutuzov responded from Berlin: "It is no surprise to me that they call my unfortunate friend a Martinist since that name is given indiscriminately to everyone. But how can it be otherwise when they don't reflect on anything, even the very word Martinist still remains a riddle which they haven't solved."[127]

Memoirist I. V. Lopukhin (1756–1816) wrote and translated numerous Masonic works. A confidant of Novikov and a fellow Rosicrucian, he founded a press in Moscow in 1783 to help shape public opinion in favor of Freemasonry.

In a letter to his brother Freemason Prince N. N. Trubetskoi, Kutuzov observed that seeking to know who the Martinists were is the same as seeking to know "the name and the qualities . . . of that frightful giant against whom the valiant knight Don Quixote so unfortunately fought." He admitted that while the putative Martinists owed their name to Saint-Martin, whose writings, Kutuzov insisted, were unfairly criticized by the "Messrs. Enlighteners," the individuals so indiscriminately defined stood for and embodied so much more than the ideas expressed by the French writer. "To put it briefly," Kutuzov continued, suddenly adopting with pride the label of Martinist,

> each land now has its Martinists, and they are nothing other than people retiring from noisy and purposeless conversations, trying to perfect the qualities impressed upon them by the Creator and, disdaining daydreaming and vanity, approximating the aim of the true man. Whether all thus named Martinists are in fact on the true path, that is a different question about which, however, loud simpletons don't want to, and are not even able to, reflect. Deny God, deceive skillfully, joke cleverly, ravage your neighbor, slander and say spiteful

things, seduce young, innocent girls—then you'll be a good and safe citizen; but abstain from all these fashionable qualities, then you will assuredly earn the name of Martinist or the most dangerous man in society.[128]

Their enemies were not the only ones tilting at windmills, however. Just as many Russians thought they saw Martinists—whom they suspected of importing revolution to Russia—lurking in every dark corner, these Freemasons exhibited similarly exaggerated powers of perception. Writing to his friends in Moscow in April 1792, Kutuzov professed knowledge that the recent "plot against holiness and order" was to a very large extent the work of the Illuminati, of whom, he asserted, "even in our fatherland there are more than a few." He warned them to be on their guard, to make sure that these "monsters" did not sneak into their ranks; he charged each brother, as a "true Freemason" and as a "good citizen and a loyal subject," to use everything at his disposal to unmask these enemies of "lawful authority and the social good." Lopukhin responded that the Moscow brothers were already well aware of the Illuminati's existence in Russia (his earlier assertion notwithstanding) and assured Kutuzov that they had already "issued all possible orders to uncover these hidden enlighteners." No one, Lopukhin concluded, was better suited to unmask them than true Freemasons.[129]

The views expressed in these letters—the confusion and ignorance about the term "Martinist," along with the belief in the existence of a dangerous band of Illuminati in Russia whom the Masons would soon reveal—should not be taken at face value, however, for the authors were well aware that they were not the only ones reading the correspondence. The Moscow postal director, I. B. Pestel', opened all their mail and made two copies of each letter—one for the governor-general of Moscow, Prince A. A. Prozorovskii, the other for Count A. A. Bezborodko in St. Petersburg, who reported directly to Catherine any important information gleaned from them. That the Masons knew of this is not surprising since the inspecting of mail (known in Russian as *perliustratsiia*) was a common practice in Russia and throughout Europe and was employed not only against suspicious individuals, but even against trusted and highly placed personages.[130] Through their correspondence, Lopukhin, Kutuzov, and their Moscow comrades, who by the mid-1780s had come to represent to society the primary Martinist group, sought to influence the authorities by presenting the "truth" about Martinism and to assure these officials that the activities of Freemasons in no way mirrored those of the French revolutionaries or of native subversives like Radishchev.[131]

They also realized that they could not overlook public opinion, which had become decidedly hostile. As Lopukhin later recalled, he and his lodge brothers had to confront more than the displeasure of Catherine and local officials at this time; they also had to respond to the "rumors and uproar of the public," whose members were growing increasingly fearful of Freemasonry and its supposed allies.[132] Russian educated society came to see the Jacobins and Freemasons, allied with the Martinists, as forming a "gang of plotters, bloodthirsty monsters, demolishers of authority, freethinkers," whereas the broad masses, largely ignorant of the Jacobins, constructed a terrifying image of the Freemasons and Martinists, whom they pictured as having "little tails."[133] In Tula in 1790, a mammoth critique of *On Errors and Truth* was published that, while not making direct reference to the events in France, accused its author of atheism and argued that the book—capable of leading its readers into "error" and, consequently, of producing "ruinous disorder, anarchy"—posed a threat to society.[134] By 1792, pamphlets and the press routinely warned of an international plot directed against the established order, and secretly led by all manner of Masons, Martinists, Encyclopedists, and other extremists.[135]

Facing this rising tide of fear, the Moscow Masons published two works in an attempt to defend their public reputation. Lopukhin composed a short piece entitled *The True Freemasons' Edifying Catechism* that he published in French and sold as a foreign text newly arrived in Russia.[136] Next, he charged his lodge brother I. P. Turgenev with writing a more developed defense that appeared anonymously in Moscow in 1790 as *Qui peut être un bon citoyen et un sujet fidèle?* (Who Can Be a Good Citizen and a Loyal Subject?) again in a deliberate attempt to disguise its origins, before going through three Russian editions by the end of the century.[137] Both works sought to counter the widespread view that the Martinists were atheists bent on undermining the political and social order. Both works highlighted the generally religious and specifically Christian nature of the order and reiterated the assertion already noted in other defenses that Freemasonry was not opposed to the established authority, and that it made its members better subjects and citizens, better sons of the native land, better Christians, and better husbands and fathers.

The question posed by the title of Turgenev's book found no easy answer. While there was little difference of opinion about the general characteristics of the good citizen and the loyal subject, there was clearly disagreement over how these qualities might best be acquired and who best embodied them. Russia's Masons remained convinced that they alone knew the path to true virtue and enlightenment that lay at the heart of such models; their critics appeared equally certain in their mistrust of the entire Masonic movement—of its members, practices, goals, and influence. This mistrust is as old as Masonry itself: fear of Freemasonry followed soon after the appearance of the first lodges in Russia. But if the fear of Masonry was constant, the way in which it expressed itself was not. Freemasonry functioned as a blank screen upon which Russians could project a variety of fears that changed with time and reflected some of the chief concerns of a given pe-

I. P. Turgenev (1752–1807) was the director of Moscow University at the turn of the century. The author of a popular defense of the Masons, Turgenev belonged to lodges in both capitals and established the Golden Wreath lodge in his Simbirsk home in 1784.

riod. The breakdown of traditional Russian social and religious customs; the baleful influence of excessive cultural borrowing; the vain desire for social distinction and its deleterious consequences; the domination of foreigners in Russia's political, cultural, and intellectual life; confusion over Russians' obligations to the state, society, and family; and the danger of social upheaval, anarchy, and revolution were the major questions that preoccupied educated Russians in the eighteenth century. Freemasonry cut across each one of these questions, acting as a lightning rod for a whole range of contentious issues and debates.

Why Masonry? What was it about the Masonic movement that made it so potent and controversial, capable of representing so many different threats at different times to different groups? A definitive answer to this question is difficult to provide. Part of the answer may lie in the large numbers of men who were attracted to the lodges and helped make Freemasonry the largest elite social movement in the entire century. In addition, the predominance of the mighty and the powerful in the Masons' ranks may well have led outsiders to exaggerate the movement's import and influence. An institution that large and that tightly interwoven with the centers of power and authority was bound to attract notice and to provoke speculation about its ultimate purpose. Also, the lodges' geographic origin, together with the Masons' uninterrupted interactions with non-Russians both inside and outside Russia, lent them an aura of unmistakable foreignness, a quality that made them suspect among many strata of society, in which there was an increasing consciousness of Russia's uniqueness (even superiority) vis-à-vis the West. But of all possible factors, undoubtedly the most important was the shroud of secrecy behind which the Freemasons gathered and with which they concealed their affairs. To most Russians, only nefarious deeds had to be hidden and only individuals ashamed of their actions concealed themselves from view. It warrants noting, however, that the distrust of Freemasonry and the primary accusations leveled against it were not unique to Russia, but largely mirrored similar patterns found throughout the Continent.[138]

The material form of the debate over Freemasonry—beginning in the middle of the century with unpublished satires and squibs copied by hand, and ending in the final decades with printed books, pamphlets, and periodicals sold in bookstores—provides us with some insight into the poorly understood dynamics of public opinion in eighteenth-century Russia. The change in form demonstrates the rapid development of the technological base necessary for the construction of the modern notion of public opinion as outlined in chapter 2 (namely, printing houses; newspapers, journals, and books; readers and booksellers) and of the shared awareness among educated Russians of the idea of public opinion. Both sides in the debate over Freemasonry saw the necessity of "going public," of using print to influence the opinions of the men and women who made up the public. The shift from manuscripts intended for an intimate circle of friends to printed books and journals able to reach a larger audience reflects a broad transformation in the country's intellectual life involving all producers of written texts.[139] This fact says much about the nature of Russian society and politics in the eighteenth century. It attests to the formation of a new source of power (viz., the public) as well as a new arena for cultural, intellectual, and political activity (viz., the sphere of print). Although the state remained the final locus of power, by the latter decades of the century the public as a site and matrix of moral authority could not be and was not ignored. Critics and defenders of the order saw the importance of winning the approval of the public as a source for the legitimization of their respective opinions.

. . .

Despite their best efforts the Moscow Masons eventually lost the battle on both fronts. In the face of mounting suspicion—even outright paranoia—among elements of society and certain state officials, Prozorovskii arrested Novikov as the recognized leader of the group in 1792. The events surrounding Novikov's arrest and imprisonment have long intrigued students of Catherinian Russia, and in spite of all sorts of theories and interpretations, no definitive explanation of this event has yet been offered. What can be unequivocally stated is that the "Novikov affair" resulted from an intricate web of factors knit together by Novikov's actions as a Freemason, and that his arrest was instigated by local officials and then presented to the empress as a fait accompli.[140] Based on Novikov's interrogation and on Catherine's ukase of 1 August 1792 sentencing him to fifteen years confinement, it is clear that Masonic activities aroused suspicion on many levels. First, the use of religious objects, symbols, and rites was seen as a direct threat to the official church; second, dealings with the Masonic authorities in Berlin—whose leaders were the political rulers of Russia's nemesis, Prussia—represented the breaking of one's oath of loyalty and the irresponsible neglect of one's duties as a subject; third, attempts to lure Grand Duke Paul into their sect—whose displeasure toward his mother and sympathies toward the Prussian court were no secret—constituted a bold challenge to Catherine's authority; and finally, the publication of works of (supposed) questionable moral worth and the ability to attract wealthy patrons to their ranks were construed as self-serving attempts to dupe the public that were unrelated to furthering the common good.[141]

Surprisingly, in light of the seriousness of the charges and the severity of the punishment, none of Novikov's comrades, several of whom were questioned by Prozorovskii in the months following Novikov's arrest, received similar treatment. Catherine merely ordered that Prince N. N. Trubetskoi, I. V. Lopukhin, and I. P. Turgenev leave Moscow for their provincial estates and desist from their previous Masonic activities. The order was even overturned in the case of Lopukhin, and he was allowed to stay in Moscow so that he might care for his aged father.[142] Two young members of Novikov's circle, M. I. Nevzorov and V. Ia. Kolokol'nikov, who had been sent abroad to study medicine, were arrested in Riga as suspected Jacobins upon their return in 1792 and detained in the Schlüsselburg fortress for questioning. Nevzorov refused to cooperate, claiming that as a student of Moscow University he would only answer to its officials, and, amazingly, he and his friend were granted their request to be turned over to the university's curator, I. I. Shuvalov, a former Mason himself.[143] No other members of Novikov's circle were interrogated or punished for their involvement.

The rest of Russia's Freemasons fared equally well, and the arrest of Novikov in no way signaled the beginning of an organized governmental campaign against the movement. Indeed, despite the measures taken against the Moscow brothers, Masons continued to meet in various towns and cities throughout the empire. In 1794, for instance, Mitau's Lodge of the Three Crowned Swords and Reval's Isis were still in operation.[144] Nevertheless, by the early 1790s the Masonic movement had run its course, and lodges throughout Russia began to shut their doors for good. When, after three decades on the throne, Catherine II died in 1796, Freemasonry as an active, organized institution had ceased to exist.

The precise role Catherine played in the closing of the lodges is less than clear. By 1790 the empress shared the popular notion that the Freemasons and similar secret bodies were responsible for the chaos that had seized France and threatened the rest of Europe. Prompted by reports of a foreigner living in Moscow with links to the Paris rebels, Catherine wrote Prozorovskii in April 1790 ordering him to close all local lodges and other secret assemblies. Around this time she also instructed General P. I. Melissino to close his St. Petersburg lodge. During the next few years Catherine was content merely to let the movement's leaders know that the time had come for the brothers to shut their lodges. Notes made in the minute books of several lodges in 1794 suggest just such a state of affairs.[145] But most of Russia's Masons did not look to the tsarina for instructions about their lodge activities because they had already ceased operations by then. The Freemasons themselves initiated the closing of many lodges, probably in response to the unsettling waves of fear, terror, and revolutionary war sweeping over Europe. Confronted by countless stories of bands of Jacobins, Illuminati, Martinists, and Freemasons furtively infiltrating country after country and spreading their infection of murder and mayhem, and by the disturbing news of the victories of France's armies and the march of revolution eastward, Russia's Masons began to doubt the stated goals of their order and thought it safer simply to stop their own activities until the world had returned to normal. Others experienced a much more profound transformation in their view of the fraternity. Johann August Starck, a onetime instructor at St. Petersburg's Peterschule and prominent figure in Russian and German Masonic circles, developed in a series of writings in the early 1790s a conspiratorial theory of the French Revolution that placed the blame squarely on the shoulders of atheist philosophers, radical Illuminati, and Freemasons. The Viennese journalist and professor Leopold Alois Hoffmann, another former Mason, devoted his energies at the end of the century to exposing the supposed evil deeds of the Freemasons, Jacobins, and Illuminati.[146]

The collapse of the movement in Russia in the 1790s paralleled developments in other parts of Europe. In the Netherlands a once vibrant Masonic community fell into a state of dormancy as many lodges closed their doors. In various German states, rulers introduced measures to curtail the activities of the Freemasons. In Austria, following mounting government pressure under the emperors Leopold II and Francis II, the beleaguered brothers were eventually hounded into disbanding in 1794; the next year nine Jacobins were executed and the death sentence was introduced for anyone caught involved with secret societies. In Switzerland all the lodges ceased operation between 1792 and 1793. In Savoy-Piedmont, Victor Amadeus III, the king of Sardinia, outlawed Freemasonry in 1794. In Portugal and Spain, the Inquisition unleashed another wave of suppression against the Freemasons. In France between 1789 and 1795, the lodges found themselves in the paradoxical (and awkward) position of being suspected by the revolution's opponents as its chief fomenters and by its most radical proponents as atavistic relics of the hated old order. "Freemasonry," to quote one authority, literally "fell apart."[147] Freemasons could no longer find a safe home anywhere in Europe.

The ascent to the throne of Paul I in the autumn of 1796 brought a reversal of fortune to the few Masons punished four years earlier. Novikov was released from prison; Trubetskoi and Lopukhin were made senators; Turgenev was appointed director of Moscow University; and Prince Repnin, who had been "demoted" to the position of governor-general of Lifland and Estland, was promoted to field marshal.[148] Paul's reign did not usher in a fresh beginning for the movement itself, however. Even though the Masons had looked to Paul to be a new patron of the order, Russia's lodges remained inactive as they had for several years. Indeed, the question of Paul's relationship to the Freemasons, like so much of the history of the order, has long been shrouded in mystery and is but one more chapter in Russian Masonry's complex, convoluted mythology. The assertion of one recent biographer notwithstanding,[149] most observers have taken Paul's membership in the order as beyond question and have gone to great lengths to justify their claims, despite the scanty evidence—supposedly a result of the purposeful destruction of documents linking the emperor to the Masons by various "interested persons" including Paul himself.[150] Intriguing as the question may be, it is of little importance for the fate of Russian Freemasonry. Even before Paul came to power, the brightest days of Russia's Masonic movement were already past.

Conclusion

In 1773, as was then common among fashionable members of society, Count Zakhar Grigor'evich Chernyshev ordered for himself an exquisite snuffbox from Petersburg craftsmen. Atop the oval box of deep blue enamel sits a miniature of Chernyshev, rimmed in gold filigree and crowned by a gently curving ribbon of silver. Below the undulating rolls of his white peruke, his languid gaze and faint hint of a smile, a bright blue sash peaks out from under Chernyshev's green coat, to which is pinned an eight-pointed silver star—both symbols of the Order of St. Andrew, imperial Russia's oldest and highest order, awarded him by Catherine II. Beneath the miniature, between its frame and the lid's edge, a cluster of Masonic symbols has been engraved: a square, a hammer, a globe, and compasses. Removing this eye-catching trinket from his pocket, Chernyshev would announce to all gathered around him his membership in the brotherhood of Freemasonry and the sense of pride and distinction with which this affiliation filled him.[1]

Chernyshev's snuffbox exhibited two sets of symbols, two spheres of social activity: one anchored in the state, one in society. While state service—the quest for high office, for the favor of the ruler, and for the power, wealth, and status that flowed from them—retained its traditional preeminence in the lives of men like Chernyshev, by the second half of the century new social spaces and practices like Freemasonry offered these men separate arenas and institutions from which they might draw a sense of purpose, meaning, and identity. Membership in the Order of St. Andrew marked Chernyshev for his exemplary service to the state, setting him apart from others as worthy of the utmost honor and respect. Membership in the order of the Freemasons marked Chernyshev and the thousands of his fellow Masons as belonging to a distinct and superior social body, one that extended beyond Russia's borders, uniting under its aegis all men of true virtue and enlightenment and directing their combined energies toward the perfection of self and society. The inclusion of a book, parchment, and quills alongside the box's Masonic symbols suggests Freemasonry's connection to more general intellectual and social developments. Being a Mason also meant being a man of learning, someone accustomed to the enlightened pursuits of educated society and familiar with the cosmopolitan culture of the European elite.

In Chernyshev's ornate snuffbox, the importance Russians attached to their identity as Masons is physically preserved and vividly displayed. I have sought to examine this identity and, in so doing, to rethink Freemasonry's place in eighteenth-century Russian society. This was a society undergoing radical transformation, including the profoundly experienced breakdown of traditional social structures and customs, the rise of an absolutist state, the growing importance of education and learning, and the rapid, wholesale borrowing of Western cultural forms that had come to dominate the lives of the elite. A pervasive, deep-seated concern with questions of personal identity, social status, and group affiliation accompanied this wrenching process as a growing body of Russians had to develop new modes of self-definition and find new ways to orient themselves within society.

Russians found in the Masonic movement a powerful basis for the cultivation of new identities and a comforting social anchor amid the perceived chaos and indeterminacy of Russian society. It gave them a sense of personal and group identity, of connectedness and community. It gave them access to specific cultural capital—as the personification and defenders of virtue and enlightenment—that accorded well with their sense of self-worth and assuaged deep anxieties over their own social standing. Membership in the lodges signaled both a separateness from the rabble and a superiority over other members of educated society. As a Mason, one belonged to the ultimate elite and shared a bond—all the more cherished because of its secrecy—with the one true nobility, that is, the nobility of the soul.

The perception of Freemasonry as an antidote to disorder, an instrument for imposing structure on a social reality of perceived formlessness, represents a difference between the way the lodges were adopted in Russia and the way they were adopted in other European countries. In western Europe the lodges appear to have fulfilled an entirely opposite need: they offered a sphere free from the deep societal divisions that made social intercourse across the traditional barriers of corporate estate, rank, and caste so difficult. In the lodge, men—and some women—could transcend the rigid hierarchies that structured society and patterned social interaction and could inhabit a space in which such distinctions were given no significance (even if only in theory or for brief moments).[2] For Russians, on the other hand, Freemasonry's great allure was in large part due to its usefulness in creating and imposing structure on society.

In a similar vein, one can also point to the greater emphasis Russian Masons placed on the lodges' function as schools for the inculcation of morals and manners. This is not to say that Masonic practice was not viewed similarly by Freemasons in other parts of Europe, but that Russians were more acutely aware of this aspect of the Masonic project and chose to emphasize it, given their heightened sense of being on the periphery of civilized Europe. By "working the rough stone," Russian Freemasons believed they were becoming more civilized and cultured, like their brothers in the West. They saw themselves as engaged in a process of self-improvement that brought them closer to the norms of the world's refined capitals and set them ever further apart from the ways of their rough and crude surroundings.

To maintain that Freemasonry's popularity in Russia can be traced in large part to the desire for distinction through the exaggeration one's own self-worth and importance (in a word, vanity) may strike some as belittling its historical significance. Liah Greenfeld, however, has argued in a recent major work that nationalism and even the modern world itself are the products of vanity: "It would be a strong statement, but no overstatement," she writes, "to say that the world in which we live was brought into being by vanity. The role of vanity—or desire for status—in social transformation has been largely underestimated."[3] As Greenfeld notes, the historical role and significance of vanity are still little appreciated and poorly understood. Nevertheless, the assertion that vanity and the desire for status lay at the core of Freemasonry's popularity should not lessen an evaluation of the movement's importance for the history of Russia. Rather, it should make clearer our understanding of Freemasonry and of Russia's past. This new perspective reveals overlooked connections between the lodges and Russian society and adds further evidence to the view of this society as one in which questions of identity and status were particularly acute.

Any attempt to understand Freemasonry must account not only for the motivations of individuals but also for the broader social landscape. The lodges were not isolated and unique institutions, but part of a complex network of new social spaces including assemblies, salons, clubs, theaters, and societies. Although concentrated in Russia's two capitals, by the final decades of the 1700s these institutions had spread to numerous provincial cities and towns throughout the empire. Together they constituted a public sphere that was the home of Russia's educated society, men and women whose learning and culture set them apart from the common people and gave them a sense of affiliation and cohesiveness. Because Russia has traditionally been viewed as a country lacking any independent society, the recognition of a Russian public sphere, or what Marc Raeff has called a "civil society," is of special importance.[4] It presents a Russian society that exhibits greater similarities with the societies of western European countries than traditional scholarly interpretation would allow. For too long Russia has been used as a foil to highlight the distinctiveness (and usually superiority) of Western nations: what makes the West different from Russia, it is often asserted, is its possession of an independent, civil society. But as Charles Taylor has cautioned, one must be careful when civil society is invoked in this manner, for neither the existence of civil society in the West nor the lack thereof in Russia is as unequivocal and evident as some would believe.[5]

Chernyshev's snuffbox neatly shows that by the second half of the eighteenth century Russians recognized the mutually constitutive distinctions of public and private, state and society. Yet, like the sets of symbols placed side by side atop this elegant object, state and society had not become fully differentiated and still remained interwoven. While such a lack of differentiation was not unique to Russia, it was more visible there than in most other parts of Europe and indicated a pervasive lack of clear social boundaries and distinctions that characterized the entire history of imperial Russia.[6] Important as this is, it should not be taken as proof of the absence of civil society in Russia or of Russia's essential difference, for in the eighteenth century the actual relations between state and society in Europe differed from country to country, as did the idea of their proper relationship. Just as there is not one universal instance of civil society to which all others must measure up or be deemed wanting, neither is there one universal idea. What united all European theorists of civil society at that time, according to Adam Seligman, was not a single, shared definition, but an overriding concern with "the problematic relationship between the private and the public, the individual and the social, public ethics and individual interests, individual passions and public concerns."[7] The words and deeds of Russia's Freemasons and their critics, as well as the affairs of the entire Russian public, show an intense preoccupation with this very same "problematic relationship," as men and women sought through their engagement with these seemingly irreconcilable oppositions to construct their own meaningful field of social action. The civil society that emerged in this setting remained a fragile construct throughout the century, as highlighted by the fate of Russia's lodges in the wake of the French Revolution. Nevertheless, it would be misleading to attach too much significance to their collapse for two reasons: first, the demise of Russian Freemasonry in the 1790s mirrored events transpiring across the entire continent and was not unique to Russia; second, the downfall of the lodges represented not a complete destruction of Russia's civil society, but its temporary constriction until the early decades of the nineteenth century.[8]

Chernyshchev died in 1784 at the height of Russia's Masonic movement. He did not live to see its demise at the end of the eighteenth century or its rebirth at the beginning of the next, although many of his fellow Freemasons did. Upon closing the lodges in the final years of Catherine's reign, Russia's brothers carefully gathered up their statutes and charters, their catechisms and minute books, their Masonic attire and lodge regalia, and put them aside for safekeeping until the day they might once again come together as brothers and practice the royal art. That day arrived sooner than many must have thought possible. Just as the new century was dawning, a handful of Masons gathered on 15 January 1800 in a small wooden house on St. Petersburg's Vasil'evskii Island. The Dying Sphinx lodge met under the strict leadership of A. F. Labzin, the famous mystic initiated as a young student into the mysteries of Freemasonry and Martinism by the Novikov circle in Moscow, whose surviving members now blessed their protégé's attempt to breathe new life into the order. Two years later, A. A. Zherebtsov founded the city's second lodge, Les Amis réunis, along French lines, and in 1805, I. V. Böber, another longtime brother, revived the Pelican of Charity lodge (founded in 1773), assisted by several of its former members. Freemasonry returned to Moscow in 1803, when the senator P. I. Golenishchev-Kutuzov and a few Rosicrucians founded a secret lodge called Neptune— the name of the former lodge at Kronstadt to which Kutuzov had been admitted as a young seaman and whose official documents its Grand Master, Admiral S. K. Greig, had given him to preserve many years before.[9]

The next decade and a half witnessed the dramatic revival of Masonic life in Russia. Once again scores of men (roughly 1,600) joined the order either as old brothers returning to the fraternity or as new brothers, members of a younger generation that had not yet come of age during Masonry's previous incarnation.[10] Just as before, the Masons of the early nineteenth century represented a variety of nationalities, confessions, and social stations. They were courtiers, high state officials, aristocrats, and nobles; they were men of letters and the arts; they were doctors and merchants. They established new lodges throughout the empire in Reval, Mitau, Poltava, and Vologda, in Kiev, Kronstadt, Tomsk, and Simbirsk, in Yamburg, Zhitomir, Nizhnii Novgorod, and Odessa.[11] As one might expect, the movement evinced many of the same traits seen in the previous century. There was the proliferation of Masonic rites and systems and Byzantine organizational structures, with all manner of higher Masonic grades, hidden bodies, and invisible Masonic superiors. There were the usual personalities, the common and the mundane along with the intriguing, the powerful, and on occasion, the dishonorable, all drawn to the lodges for their own particular reasons—some noble, some less so. There were the nagging suspicions between the native Russians and such foreign brothers as the German Ignatius Fessler (who was openly hated) that highlighted the gap between lofty Masonic ideals and petty everyday reality and threatened to tear apart the order. There was the undying fascination with mysticism and magic, with alchemy, Rosicrucianism, and the secret sciences in general. There was the continued preoccupation with "working the rough stone," and there was the brothers' self-conscious rehearsing of their moral superiority, gestures that were thrown into bold relief by the persistent rumors of the Masons' licentiousness and corruption.

Yet in one important way Freemasonry in the early nineteenth century differed from its predecessor—namely, in its relationship to the state. In the new postrevolutionary world order, in which the link between the lodges and the conspiratorial forces of radicalism had become permanently fixed in the public imagination, Freemasons could no longer assemble without giving thought to how their existence would be perceived by the authorities. And so from the earliest years of the century, Russia's new Masonic leaders made certain that word of their activities was unambiguously communicated to the highest official levels. According to Masonic legend, Böber was granted a private audience with Alexander I in 1803, at which he convinced the young tsar of the order's benign influence and persuaded him to lift the ban against the lodges. Supposedly, Alexander not only acquiesced to Böber's request, but even asked to be initiated into the fraternity. Whether Alexander joined the order at that time or several years later under different circumstances is a matter of debate. It is beyond question, however, that throughout Alexander's reign the Russian government made a much greater effort to monitor the Masons' activities than it had in the eighteenth century. Thus, in 1810, during a thorough police investigation of the lodges, conducted with the cooperation of most of the movement's leaders, the minister of police, A. D. Balashov, himself a member of Les Amis réunis and an initiate into the higher grades, made the unusual offer of state protection for the order if the brothers would submit to complete police subordination. Faced with the dilemma of choosing between the loss of autonomy in return for official patronage, on one hand, and the rejection of explicit government favor for the maintenance of complete freedom, on the other, Russia's Freemasons opted for the latter.[12] It was a decision that many brothers probably came to regret.

The Masons often referred to the order's revival under Alexander I as a "golden age." Yet brilliant as it undeniably was, the rebirth of Russian Freemasonry turned out to be as brief as it was bright, and by 1820 the outlines of the forces—both domestic and foreign, both internal and external to the lodges—that would eventually combine to destroy this second flowering of the movement could be seen coming into view. The rebellions of 1820 in Spain and the Italian states, followed by the mutiny of the elite Semenovksii guard regiment, were interpreted by Alexander as symptoms of a much broader revolutionary threat facing all of Europe. He had been led to this perception in large part by the Austrian statesman Clemens von Metternich, who believed Europe was locked in a manichaean struggle between legitimate, god-sanctioned political and social authority and subversive, irreligious radicalism. In a letter to the tsar in 1820, Metternich described the secret societies as the chief agent responsible for spreading this poison. These societies formed, he wrote, "a real power, all the more dangerous as it works in the dark, undermining all parts of the social body, and depositing everywhere the seeds of a moral gangrene which is not slow to develop and increase." That same year at the Congress of Troppau, when Metternich had the opportunity to meet privately with the tsar, he desperately sought to prove to Alexander the danger the secret societies represented and the need to take action against them. Metternich achieved his goal, and by the time Alexander departed he was convinced of the threat posed by secret societies and of the necessity to counteract their influence.[13]

More calls for action against the forces of instability greeted the tsar upon his return to Russia. Surprisingly, one of the loudest voices raised against the order belonged to one of its leaders—lieutenant-general E. A. Kushelev, elected Grand Deputy Master of the Grand Lodge Astreia in 1820. An older Mason with the sensibilities of an earlier, bygone era, Kushelev expressed grave concerns about the present state of the Masonic movement in a series of memoranda to the tsar. It is a double-edged sword, he wrote, that in the hands of "true" Masons is put in the service of virtue, religion, humanity, and the state, but that if seized by false Masons can easily become a powerful weapon for atheism and freethinking. In Russia, as in many parts of Europe, Freemasonry had recently lost its original purpose and the lodges had become "nests of discord, self-will, [and] unruly, riotous conduct." Kushelev proposed two courses of action: either the lodges be returned to their original, nonpolitical spirit and reorganized into a unified body under central leadership and government supervision, or they be disbanded and outlawed. The need for swift action, however, was imperative. Without it, he prophesied, "nothing can be expected [from the lodges] other than disastrous consequences such as have already come and are continually coming to light in other European states, destroying ancient and wise governments and shaking even the thrones of monarchs, the fathers and well-doers of people, plunging the people themselves into incalculable calamities."[14]

Soon after receiving Kushelev's warning, Alexander took action. On 1 August 1822, Count V. P. Kochubei, the Minister of Internal Affairs, issued a rescript in the name of the emperor instructing all Masonic lodges to close. On 10 August, Russia's Masons convened to listen to the decree and to sign a pledge that they would never again establish this or any other secret society. After signing the oath and settling their remaining affairs, the brothers closed their lodges for the last time; there were no reports of disobedience or resistance. While some brothers supposedly met the destruction of the order with indifference, others claimed that nothing could ever break the tie that bound them, and still others talked of continuing to meet—though in secret—at their dachas.[15]

In the following decades, Masonic life all but disappeared from Russia. In 1826, Alexander's ban was repeated by his successor, Nicholas I, in the wake of the Decembrist Uprising, many of whose participants belonged to the lodges that were seen in some circles as connected with the revolt. Not all Russian Freemasons renounced their former ways, however, and there are reports of small groups of brothers assembling, usually more as circles than as proper lodges, in the 1830s. It was rumored that even into the 1850s a lodge secretly met on Moscow's Polianka Street under the direction of a well-known local priest. And as late as 1870, the police were receiving denunciations of supposed Masonic gatherings, most of which appear to have been fabrications of overly suspicious and impressionable minds.[16] During the second half of the century, Freemasonry survived among Russian expatriates who took an active role in Masonic life abroad. It was from these émigré circles, primarily in France, that a second revival of Freemasonry was launched in the beginning years of the twentieth century. Little is known about this chapter of Russia's Masonic history, largely due to a lack of sources and to historians' unease with a topic tainted by its association with right-wing Russian groups intent on cultivating a myth about the Judeo-Masonic origins of the Russian Revolution. Whatever the precise role Masons played in the downfall of tsarism (a role that even some Russian Masons believed not insignificant), Freemasonry of the early twentieth century was first and foremost a political phenomenon viewed by its leaders as a useful instrument for uniting and directing the (mostly liberal) forces opposed to autocracy. Not surprisingly then, with the collapse of the Romanov dynasty the new political Freemasonry disappeared along with its very raison d'être.[17] Freemasonry's fate under Soviet rule is largely a matter of speculation. Sketchy evidence suggests that small groups of Masons continued to meet throughout the 1920s only to be uncovered and subsequently destroyed in the successive waves of repression and terror of the 1930s. The collapse of the Soviet Union has created the proper conditions for a third Masonic revival, and once more lodges are being established in Russia.[18] Time will tell whether this most recent incarnation proves more lasting than its predecessors.

Notes

List of Abbreviations

ch.	*chast'* [part]
d.	*delo (dela)* [file (files)]
ES	*Entsiklopedicheskii slovar' Brokgauz-Efron*
e./x.	*edinitsa (edinitsy) khraneniia* [unit (units) of storage]
f.	*fond* [(archival) collection]
l. (ll.)	*list* [leaf (leaves)]
ob.	*oborot* [back/reverse side]
op.	*opis'* [inventory]
OPI GIM	Otdel pis'mennykh istochnikov Gosudarstvennogo istoricheskogo muzeia [Division of Written Sources, State Historical Museum]
OR RGB	Otdel rukopisei Rossiiskoi gosudarstvennoi biblioteki [Manuscript Division, Russian State Library]
OR RNB	Otdel rukopisei Rossiiskoi natsional'noi biblioteki [Manuscript Division, Russian National Library]
ORRK BAN	Otdel rukopisnoi i redkoi knigi Biblioteki Akademii nauk [Division of Manuscripts and Rare Books, Library of the Russian Academy of Sciences]
PD	Rukopisnyi otdel, Institut russkoi literatury (Pushkinskii dom) Rossiiskoi Akademii nauk [Manuscript Division, Institute of Russian Literature (Pushkin House), Russian Academy of Sciences]
PSZ	*Polnoe sobranie zakonov Rossiiskoi imperii*
RA	*Russkii arkhiv*

RS   *Russkaia starina*

RGADA   Rossiiskii gosudarstvennyi arkhiv drevnikh aktov [Russian State
              Archive of Early Acts]

RGALI   Rossiiskii gosudarstvennyi arkhiv literatury i iskusstva [Russian State
              Archive of Literature and Art]

SK   *Svodnyi katalog russkoi knigi grazhdanskoi pechati XVIII veka, 1725–1800*

TsKhIDK   Tsentr khraneniia istoriko-dokumental'nykh kollektsii [Center for the
              Preservation of Historico-Documentary Collections]

TsNB-KRK
GE   Tsentral'naia nauchnaia biblioteka—Kabinet redkoi knigi Gosu-
              darstvennogo Ermitazha [Central Scientific Library—Rare Book
              Room, State Hermitage]

INTRODUCTION

1. W. Gareth Jones, *Nikolay Novikov, Enlightener of Russia* (Cambridge, 1984), 206–15; M. N. Longinov, *Novikov i moskovskie martinisty* (Moscow, 1867), 313–23.

2. Sergei Nekrasov, *Apostol dobra. Povestvovanie o N. I. Novikove* (Moscow, 1994); Jones, *Nikolay Novikov, Enlightener of Russia.*

3. A leading authority writes: "Probably nowhere in Europe did Freemasonry play so big a part in the development of the cultural life of three or four generations as it did in Russia." Isabel de Madariaga, *Russia in the Age of Catherine the Great* (New Haven, 1981), 521.

4. See Liah Greenfeld, "The Scythian Rome," chap. 3 in *Nationalism: Five Roads to Modernity* (Cambridge, Mass., 1992); Pavel Miliukov, *Ocherki po istorii russkoi kul'-tury*, vol. 3 (Paris, 1930), 227–30; Elise Kimerling Wirtschafter, *Social Identity in Imperial Russia* (DeKalb, Ill., 1997).

5. The expression comes from Marc Raeff, *The Well-Ordered Police State: Social and Institutional Change through Law in the Germanies and Russia, 1600–1800* (New Haven, 1983).

6. V. Bogoliubov writes in *N. I. Novikov i ego vremia* (Moscow, 1916): "Catherine's sentence against Novikov was a sentence against society's attempt at independent self-determination which by its very essence was intolerable to absolutist power. This view, one might say, has achieved general recognition," pp. 454, 457–59. See also A. N. Pypin, *Russkoe masonstvo. XVIII i pervaia chetvert' XIX v.* (Petrograd, 1916); G. P. Makogonenko, *Nikolai Novikov i russkoe prosveshchenie XVIII veka* (Moscow-Leningrad, 1951).

7. See J. M. Roberts, *The Mythology of the Secret Societies* (London, 1972).

8. Frances A. Yates, *The Rosicrucian Enlightenment* (London, 1972), 209.

9. This discussion is based on David Stevenson's *The Origins of Freemasonry: Scotland's Century, 1590–1710* (Cambridge, 1988).

10. Stevenson, "The Renaissance Contribution," chap. 5 in *Origins of Freemasonry.*

11. Frances Yates first suggested the significance of late Renaissance intellectual currents for the rise of Freemasonry over thirty years ago in *Giordano Bruno and*

the Hermetic Tradition (Chicago, 1964), 274. Stevenson's study largely validates Yates's hypothesis.

12. Stevenson, *Origins of Freemasonry*, 9–12.

13. Margaret C. Jacob, *Living the Enlightenment: Freemasonry and Politics in Eighteenth-Century Europe* (New York, 1991), 32–35; Stevenson, *Origins of Freemasonry*, 216–31; Steven C. Bullock, *Revolutionary Brotherhood: Freemasonry and the Transformation of the American Social Order, 1730–1840* (Chapel Hill, N.C., 1996), 42–43; Roberts, *Mythology*, 25–26.

14. See the excellent discussion of early English Masonry in Bullock, "Newton and Necromancy: The Creation of the Masonic Fraternity," chap. 1 in *Revolutionary Brotherhood*. On early Enlightenment sociability and the discourse of politeness in England, see G. J. Barker-Benfield, *The Culture of Sensibility: Sex and Society in Eighteenth-Century Britain* (Chicago, 1992); Lawrence E. Klein, *Shaftesbury and the Culture of Politeness: Moral Discourse and Cultural Politics in Early Eighteenth-Century England* (Cambridge, 1994).

15. Robert Freke Gould, *The History of Freemasonry*, vol. 4 (New York, 1889), 7–15; Jacob, *Living the Enlightenment*, 16, 74, 89–90; Roberts, *Mythology*, 64.

16. Jacob, *Living the Enlightenment*, 3, 73; Roberts, *Mythology*, 65–66.

17. Roger Chartier, *The Cultural Origins of the French Revolution*, trans. Lydia G. Cochrane (Durham, N.C., 1991), 162–63; Jacob, *Living the Enlightenment*, 205; Daniel Roche, "Literarische und geheime Gesellschaftsbildung im vorrevolutionären Frankreich: Akademien und Logen," in *Lesegesellschaften und bürgerliche Emanzipation: Ein europäischer Vergleich*, ed. Otto Dann (Munich, 1981), 182.

18. Chartier, *Cultural Origins*, 164–65; Jacob, *Living the Enlightenment*, 8–9, 16, 205.

19. Richard van Dülmen, *Die Gesellschaft der Aufklärer: Zur bürgerlichen Emanzipation und aufklärerischen Kultur in Deutschland* (Frankfurt a. M., 1986), 55–66; Rudolf Vierhaus, "Aufklärung und Freimaurerei in Deutschland," in *Freimaurer und Geheimbünde im 18. Jahrhundert in Mitteleuropa*, ed. Helmut Reinalter (Frankfurt a. M., 1983), 115–39.

20. Gould, *History of Freemasonry*, 1–7; Roberts, *Mythology*, 64.

21. Gould, *History of Freemasonry*, 104–5, 112–13, 117–19; Jacob, *Living the Enlightenment*, 27, 73; Roberts, *Mythology*, 68–76.

22. Gould, *History of Freemasonry*, 95–103.

23. Ernest Krivanec, "Die Anfänge der Freimaurerei in Österreich," in Reinalter, *Freimaurer und Geheimbünde*, 177–92; R. William Weisberger, *Speculative Freemasonry and the Enlightenment: A Study of the Craft in London, Paris, Prague, and Vienna*, East European Monographs, 367 (New York, 1993), 109.

24. Reinalter, "Die Freimaurerei zwischen Josephinismus und frühfranziszeischer Reaktion. Zur gesellschaftlichen Rolle und zum indirekt politischen Einfluß der Geheimbünde im 18. Jahrhundert," in Reinalter, *Freimaurer und Geheimbünde*, 35–84.

25. Bullock, *Revolutionary Brotherhood*, 46; Gould, *History of Freemasonry*, 125–49.

26. Bullock, *Revolutionary Brotherhood*, 85–90; Jacob, *Living the Enlightenment*, 59–62.

27. See Eric Hobsbawm, "Introduction: Inventing Traditions," in *The Invention of Tradition*, ed. Eric Hobsbawm and Terence Ranger (Cambridge, 1983), 6, 8.

28. François Furet, *Interpreting the French Revolution*, trans. Elborg Forster

(Cambridge, 1981), 164, 175–76, 178. Cochin's chief writings on the revolution appeared posthumously: *Les Sociétés de Pensée et la révolution en Bretagne* (1925); *Les Sociétés de Pensée et la démocratie moderne* (1921); and *La révolution et la libre pensée* (1924).

29. Furet, *Interpreting the French Revolution*, 187.

30. Ibid., 180.

31. This work was originally published in Germany in 1959; an English translation appeared from MIT Press in 1988.

32. Jacob, *Living the Enlightenment*, 20.

33. For a succint introduction to the problem, see Dorinda Outram, "What Is Enlightenment?" chap. 1 in *The Enlightenment* (Cambridge, 1995).

34. Robert Darnton, "George Washington's False Teeth," *New York Review of Books* 44, no. 5 (27 March 1997): 34–38.

35. Ibid., 34.

36. See Vierhaus, "Aufklärung und Freimaurerei in Deutschland," in Reinalter, *Freimaurer und Geheimbünde*, 115–39.

37. Jacob, *Living the Enlightenment*, 18.

38. Ibid., 119.

1. Life in the Lodges

1. Unless otherwise noted, the information presented in this section comes from A. I. Serkov, "Rossiiskoe masonstvo: Entsiklopedicheskii slovar', vol. 1, Masony v Rossii (1731–1799)," unpublished manuscript, Moscow, n.d. This manuscript, the product of approximately seven years of research, much of it drawing on previously unknown archival materials, represents the most complete source for information on Russia's lodges and Freemasons. Mr. Serkov was kind enough to allow me to make full use of this remarkable work. A more readily available source on the Masons is Tatiana Bakounine, *Le répertoire biographique des francs-maçons russes (XVIIe et XIXe siècles)*, Collection historique de l'Institut d'Etudes Slaves, 19 (Paris, 1967). On the early years of Masonry in Russia, see A. G. Cross, "British Freemasons in Russia during the Reign of Catherine the Great," *Oxford Slavonic Papers*, n.s., 4 (1971): 43–49; G. V. Vernadskii, *Russkoe masonstvo v tsarstvovanie Ekateriny II* (Petrograd, 1917), 1–10. For an opposing, and unconvincing, perspective that attempts to show a German origin for Freemasonry in Russia, see Ernst Friedrichs, *Geschichte der einstigen Maurerei in Rußland* (Berlin, 1904).

2. Keith's cousin, John, earl of Kintore, was the Grand Master of English Freemasonry. Anthony Cross, *By the Banks of the Neva: Chapters from the Lives and Careers of the British in Eighteenth-Century Russia* (Cambridge, 1997), 28–29; Vernadskii, *Russkoe masonstvo*, 4. Keith was remembered by Russia's Freemasons in a song supposedly sung in their lodges during the reign of Elizabeth. See S. V. Eshevskii, *Sochineniia po russkoi istorii* (Moscow, 1900), 189, 191 n.

3. Ernest Krivanec, "Die Anfänge der Freimaurerei in Österreich," in *Freimaurer und Geheimbünde im 18. Jahrhundert in Mitteleuropa*, ed. Helmut Reinalter (Frankfurt a. M., 1983), 180. Serkov, "Rossiiskoe masonstvo," 21–22, 527, 755, 1090.

4. Serkov, "Rossiiskoe masonstvo," 796–99, 895–1019, 1027–67.

5. Ibid., 847–92, 985–1019.

6. Ibid., 163–67, 845; Vernadskii, *Russkoe masonstvo*, 18; S. N. Glinka, *Zapiski* in

Zolotoi vek Ekateriny Velikoi: Vospominaniia, ed. V. M. Bokova and N. I. Tsimbaev (Moscow, 1996), 34–35. According to Serkov there were eight military lodges.

7. Friedrichs, *Geschichte,* 43, 88; Serkov, "Rossiiskoe masonstvo," 835, 893, 1070–71, 1073.

8. Tira Sokolovskaia, *Russkoe masonstvo i ego znachenie v istorii obshchestvennogo dvizheniia (XVIII i pervaia chetvert' XIX stoletiia)* (St. Petersburg, [1907]), 162; Serkov, "Rossiiskoe masonstvo," 862, 879.

9. Serkov, "Rossiiskoe masonstvo," 961; Vernadskii, *Russkoe masonstvo,* 22, 63.

10. Friedrichs, *Geschichte,* 18; Vernadskii, *Russkoe masonstvo,* 9.

11. Marie Daniel Bourrée, baron de Corberon, *Un diplomate français à la cour de Catherine II, 1775–1780: Journal intime,* ed. L. H. Lablande (Paris, 1901), 2:139.

12. Serkov, "Rossiiskoe masonstvo," 895, 901, 918, 940, 961, 1013.

13. Ibid., 1070, 1072.

14. OR RNB, f. 550, F.III.129, l. 1. Unless otherwise noted, all translations were done by the author.

15. Serkov, "Rossiiskoe masonstvo," 838–89. Tira Sokolovskaia discusses the history of the Neptune lodge in "O masonstve v prezhnem russkom flote," *More* 8 (1907): 216–53.

16. Vernadskii, *Russkoe masonstvo,* 84–85. Note: Vernadskii lists two different years for this event—in the text he writes 4 March 1786, yet a footnote on p. 85 says 4 March 1784.

17. Ibid., 10–12; Nicholas V. Riasanovsky, *A History of Russia,* 4th ed. (New York, 1984), 294–95. Serkov's manuscript lists 3,093 Freemasons active in the eighteenth century. I have drawn almost exclusively on this source for the data in this section.

18. Serkov has identified 2,020 of the 3,093 Freemasons by occupation or profession. Therefore, these 1,100 Masons represent approximately 35 percent of the total Masonic population and 54 percent of the 2,020 Freemasons with known occupations.

19. John LeDonne, *Ruling Russia: Politics and Administration in the Age of Absolutism, 1762–1796* (Princeton, 1984), 5. Other families that should be added to this list include the Chernyshevs, Iusupovs, Kurakins, Panins, and Razumovskiis.

20. Nikita Panin was the principal architect of Russia's foreign policy during the early decades of Catherine II's rule and tutored the Grand Duke Paul. Panin's brother, Petr, led the forces that put down the Pugachev Rebellion (1773–1774) and, along with Nikita, served in the empress's nine-member Council; I. G. Chernyshev was, among other things, vice-president and later president of the Naval College and a senator; Z. G. Chernyshev was president of the College of War, a field marshal, and governor-general of Polotsk and Mogilev provinces; K. G. Razumovskii was president of the Academy of Sciences, hetman of Ukraine, and a field marshal; Prince A. B. Kurakin headed the Senate's Third Department of State Revenues and became Procurator-General under Paul I.

21. LeDonne, *Ruling Russia,* 14.

22. Ibid. On the degree to which the tsarist state apparatus and court were filled with Freemasons, see also Sokolovskaia, *Russkoe masonstvo,* 160–66; Vernadskii, *Russkoe masonstvo,* 86–90.

23. A conservative estimate of 50 percent state employment for all doctors,

pharmacists, teachers, professors, musicians, actors, surveyors, and so on puts the total for those active in state service at approximately 1,240, or 41 percent of all Freemasons and 61 percent of Freemasons with known occupations.

24. Merchants, lawyers, bankers, manufacturers, innkeepers, and bookkeepers account for almost 17 percent of the 2,020 Freemasons identified by occupation. Together with the 54 percent active in state service they represent 71 percent of these 2,020 Masons.

25. David L. Ransel, *The Politics of Catherinian Russia: The Panin Party* (New Haven, 1975), 111–13, 255–61; Serkov, "Rossiiskoe masonstvo," 443–46.

26. See Cross, "British Freemasons," 62–64; Serkov, "Rossiiskoe masonstvo," 940–46.

27. Serkov, "Rossiiskoe masonstvo," 961–1011; Vernadskii, *Russkoe masonstvo*, 17.

28. OPI GIM, f. 17, op. 2, d. 406, ll. 5–5ob.

29. Cross, "British Freemasons," 63; Ransel, *Politics of Catherinian Russia*, 238; Serkov, "Rossiiskoe masonstvo," 940–46.

30. Vernadskii, *Russkoe masonstvo*, 12–13; OPI GIM, f. 17, op. 2, d. 406, l. 82.

31. Vernadskii, *Russkoe masonstvo*, 13; OPI GIM, f. 17, op. 2, d. 406, ll. 67–67ob; and d. 407, l. 89.

32. Anthony Cross has written that another lodge of British operative Masons may have met in Petrozavodsk in the 1790s. Cross, "British Freemasons," 60–62; idem, *By the Banks of the Neva*, 245–46, 253; M. N. Longinov, *Novikov i moskovskie martinisty* (Moscow, 1867), 95; Vernadskii, *Russkoe masonstvo*, 13; Serkov, "Rossiiskoe masonstvo," 866–69, 888–89, 1019–20. Serkov lists a single lodge active in Petrozavodsk in the 1790s, but this does not appear to be related to the one mentioned by Cross.

33. [Johann Philipp Wegelin], "Pis'mo neizvestnago litsa o Moskovskom masonstve XVIII veka," *RA*, bk. 1 (1874): 1032. Based on the contents of this anonymous letter and the entry on the Three Banners in Serkov's biographical dictionary, it is possible to conclude that its author was Johann Philipp Wegelin, a graduate of Moscow University, teacher, and later author of popular primers. Serkov, "Rossisskoe masonstvo," 271–72.

34. Georg Reinbeck, *Travels from St. Petersburgh through Moscow, Grodno, Warsaw, Breslaw, etc. to Germany, in the Year 1805*, in *A Collection of Modern and Contemporary Voyages and Travels*, ed. Richard Phillips, vol. 6 (London, 1807), 128.

35. OR RGB, f. 147, No. 5 /M. 1882/, ll. 47–48.

36. Serkov, "Rossiiskoe masonstvo," 612, 846–47.

37. The discussion of adoption lodges comes from Janet M. Burke, "Freemasonry, Friendship and Noblewomen: The Role of the Secret Society in Bringing Enlightenment to Pre-Revolutionary Women Elites," *History of European Ideas* 10, no. 3 (1989): 283–93; Janet M. Burke, "French Women Freemasons into the Age of Charity" (paper presented at the annual meeting of the American Historical Association, Seattle, January 1998); Janet M. Burke and Margaret C. Jacob, "French Freemasonry, Women, and Feminist Scholarship," *Journal of Modern History* 68 (September 1996): 513–49. See also the excellent discussion in Jacob, "Freemasonry, Women, and the Paradox of the Enlightenment," chap. 5 in *Living the Enlightenment*. For a more negative assessment of the relationship between women and Freemasonry, see Dena Goodman, *The Republic of Letters: A Cultural History of the French Enlightenment* (Ithaca, N.Y., 1994), 253–59.

38. Serkov, "Rossiiskoe masonstvo," 806–9, 835, 846. Serkov incorrectly calls the adoption lodge in Mitau Three Crowned Swords (the name of its brother lodge) and dates its origins to the spring of 1780. The correct lodge name and date are indicated on the diploma given to Agnese Eliesabeth de Medem [*sic*] by Grand Master Cagliostro when she was initiated into the degree of Scottish Mistress on 27 May 1779. TsKhIDK, f. 1412, op. 1, d. 5299. On Cagliostro in Courland, see V. Zotov, "Kaliostro, ego zhizn' i prebyvanie v Rossii," *RS* 12 (January 1875): 59–63. For a general overview of Cagliostro's fantastic life, see Roberto Gervaso, *Cagliostro: A Biography*, trans. Cormac Ó Cuilleanáin (London, 1974).

39. Baron Karl-Heinrich Heyking, "Vospominaniia senatora barona Karla Geikinga," *RS* 91 (September 1897): 532–35. The correspondence between Heyking's Warsaw lodge Catherine à l'Etoile du Nord and Berlin's Royal York de l'Amitié, which includes many letters penned by Heyking, is in TsKhIDK, f. 1412, op. 1, d. 4860.

40. John Robison, *Proofs of a Conspiracy against all the Religions and Governments of Europe, Carried on in the Secret Meetings of Free Masons, Illuminati, and Reading Societies*, 3rd ed. (Edinburgh, 1798), 3. For more on Robison, see Cross, *By the Banks of the Neva*, 30, 154, 258, 400 n. 175.

41. This statement is based on what we know of the lodges Urania, Nine Muses, and Bellona in St. Petersburg and the Golden Key in Perm'. On the Petersburg lodges, see Serkov, "Rossiiskoe masonstvo," 898, 940; Vernadskii, *Russkoe masonstvo*, 22 n. 3; OPI GIM, f. 17, op. 2, dd. 406, 407. On the Golden Key lodge, see the official minutes in OR RNB, f. 550, F.III.129.

42. The guidelines for this latter type of conference lodge are laid out in one especially intricate set of Masonic bylaws. See OR RGB, f. 14, No. 2, ll. 56–56ob.

43. Ibid., ll. 58–58ob.

44. OR RNB, f. 550, F.III.110, l. 12; A. N. Pypin, *Russkoe masonstvo. XVIII i pervaia chetvert' XIX v.* (Petrograd, 1916), 129–30 n. 1, 162 n. 1. See also Tira Sokolovskaia, "Traurnaia lozha u masonov (K istorii masonstva v Rossii)," *More* 6 (1906): 205–11.

45. OR RNB, f. 550, F.III.111, ll. 20ob–21ob. For a general description of the banquet that draws heavily on early nineteenth-century practices, see Tira Sokolvskaia, "Obriadnost' vol'nykh kamenshchikov," in *Masonstvo v ego proshlom i nastoiashchem*, ed. S. P. Mel'gunov and N. P. Sidorov (1914–1915; reprint, Moscow, 1991), 2:111–17. The rules used during the banquets of Kronstadt's Neptune lodge are in RGALI, f. 442, op. 2, d. 6.

46. OPI GIM, f. 17, op. 2, d. 406, ll. 43–44ob.

47. In addition to St. John the Baptist Day, each lodge held its own yearly celebration, again with an especially festive banquet, to mark the founding of the lodge. See Tira Sokolovskaia, "Ioannov den'—masonskii prazdnik," *More* 23–24 (1906): 846–63.

48. The starting time for lodge meetings was set by the General Laws used by many lodges. See, for example, those of the Rising Sun lodge in OR RNB, f. 550, F.III.110, ll. 9ob–10. The minutes of both the Golden Key lodge in Perm' and Vologda's Theoretical Degree show them beginning their meetings at either 5:00 or 6:00 P.M. Ibid., F.III.129; OR RGB, f. 147, No. 293 /M. 2167/. The example of the adornments comes from the minutes of the Golden Key lodge. OR RNB, f. 550,

F.III.129, l. 3ob. On Urania's cello, see OPI GIM, f. 17, op. 2, d. 406, l. 22.

49. OR RNB, f. 550, F.III.110, l. 9ob; F.III.111, l. 31ob.

50. Variations of this pattern include several lodges—for example, Kronstadt's Neptune—which also had a Deputy Master among their officials, or the Urania lodge, which had two Stewards in addition to the standard seven officials. The officials' duties are outlined in the minutes of the Urania lodge. OPI GIM, f. 17, op. 2, d. 406, ll. 7–8ob. See also Serkov, "Rossiiskoe masonstvo," 81–82.

51. OR RNB, f. 550, F.III.110, l. 9ob; F.III.111, ll. 17ob–20. On the Neptune lodge, see Tira Sokolovskaia, "O masonstve v prezhnem russkom flote," 216, 242.

52. The documents are in OR RNB, f. 550, F.III.110, F.III.111, F.III.112, F.III.113. They contain the Common Institutions of the Freemasons, the Exegesis of the Statutes or Regulations of the Freemasons, the General Laws, and the complete acts for the constituent three lower levels. The General Laws discussed below are in F.III.110. Other ceremonial documents belonging to the Rising Sun (for example, questions put to the various lodge officials during initiations) are in OPI GIM, f. 398, d. 17.

53. Law §42.

54. Law §39.

55. Laws §§9, 14, 16, 19, 20, 21, 22, 30, 31.

56. Laws §§20, 23, 26.

57. Law §11.

58. Law §16.

59. Laws §§3, 27, 18, 5, 23, 20, 31, 29, 33.

60. Laws §§47, 34. See also §§9, 10, 23, 29, 30, 31, 34, 47.

61. Law §50.

62. Law §15.

63. OPI GIM, f. 17, op. 2, d. 406, ll. 24, 43ob–44, 72–72ob; Vernadskii, *Russkoe masonstvo*, 64.

64. P. Pekarskii, *Dopolneniia k istorii masonstva v Rossii XVIII stoletiia* (St. Petersburg, 1869), 122–23; Vernadskii, *Russkoe masonstvo*, 64; OR RGB, f. 147, No. 5 /M. 1882/, l. 47ob. A list of the members of the Neptune lodge from 1781 includes the names of two brothers listed as being expelled, although no reasons are cited. Sokolovskaia, "O masonstve v prezhnem russkom flote," 242.

65. Pekarskii, *Dopolneniia*, 121–22.

66. [Wegelin], "Pis'mo neizvestnago litsa," 1036–37.

67. OR RNB, f. 550, Q.III.153, l. 2; F.III.111, ll. 11–11ob, 22ob. The specific sources in F.III.111 are the Brief Outline of the Order's Goal; the order's seven duties; and the Statutes or Regulations of the Freemasons.

68. *Slovar' Akademii Rossiiskoi*, pt. 2, 1790, s.v. "Dobrodetel'."

69. W. Gareth Jones, *Nikolay Novikov, Enlightener of Russia* (Cambridge, 1984), 105–7. For a succinct discussion of the concept of "mœurs" in the eighteenth century, see *The Blackwell Companion to the Enlightenment*, s.v. "Mœurs."

70. *Slovar' Akademii Rossiiskoi*, pt. 4, 1793, s.v. "Nrav."

71. Ibid., s.v. "Nravouchenie."

72. OR RNB, f. 550, F.III.111, l. 5ob.

73. Ibid., F.III.112, l. 6ob; F.III.110, ll. 3, 8ob–9.

74. Ibid., F.III.113, l. 6ob.

75. The documents sent to the Rising Sun lodge of Kazan' include both the

Statutes or Regulations and the Exegesis. See ibid., F.III.111, ll. 22ob–24; F.III.110, ll. 3–9. A copy of the former text was published in *Iz bumag mitropolita moskovskago Platona* (Moscow, 1882), 19–21. A copy of the latter, although mislabeled as simply the Statutes of the Freemasons, was published by Tira Sokolovskaia in *More* 25–26 (1907): 769–80. Also note that the document published by Sokolovskaia predates 1787, the date listed in its title, having been in use as early as 1777 among the numerous lodges operating under I. P. Elagin. Vernadskii, *Russkoe masonstvo*, xi–xii.

76. OR RNB, f. 550, F.III.111, l. 22ob.

77. Ibid., F.III.111, l. 23.

78. Ibid., F.III.110, ll. 4ob–5.

79. Ibid., F.III.111, ll. 23–23ob.

80. Ibid., F.III.110, ll. 7ob–8ob; F.III.111, l. 24.

81. Ibid., F.III.110, l. 7.

82. *Iz bumag mitropolita*, 6; OR RNB, f. 550, F.III.111, l. 13ob.

83. OPI GIM, f. 398, d. 17, ll. 1–1ob. See also OR RNB, f. 550, F.III.111, ll. 29–29ob.

84. *Iz bumag mitropolita*, 6, 7.

85. OPI GIM, f. 398, d. 16, l. 9. In a different text taken from the Rising Sun lodge, the Master asks the Senior Warden what a Freemason is, to which the latter responds: "A Freemason is a free man trying to master his inclinations and to subject his will to the laws of enlightened thought." Ibid., d. 17, l. 1.

86. This notion is suggested by the occasional usage of "*svobodnyi kamenshchik*," as opposed to the other common term "*vol'nyi* kamenshchik." OR RNB, f. 550, F.III.112, ll. 6ob–7. In a speech to the brothers of the Golden Key lodge on 15 July 1783, I. I. Panaev contrasted those individuals enslaved by their drives and passions to the Freemasons, who had "freed" themselves from such emotions. According to Panaev, the title "Freemason" referred to the "freedom of our souls" from these harmful forces. Ibid., F.III.129, ll. 31ob–33ob.

87. Ibid., F.III.47, l. 35.

88. Ibid., F.III.59, ll. 10ob–11ob.

89. E. Tarasov, "K istorii russkago obshchestva vtoroi poloviny XVIII stoletiia: Mason I. P. Turgenev." *Zhurnal Ministerstva narodnago prosveshcheniia*, n.s., 51, no. 6, section 3 (1914): 158.

90. Ibid., 159.

91. Each of the three lower Masonic degrees—as well as some of the higher degrees—had its own carpet with different signs and symbols meant to help teach the appropriate moral lessons for that particular degree. See Tira Sokolovskaia, "Masonskie kovry (Stranichka iz istorii masonskoi simvoliki)," *More* 13–14 (1907): 417–34. Note, however, that the pictures of the Masonic carpets in this article are incorrectly labeled: carpets 1 and 3 are Master degree carpets; carpet 2 is a Fellow Craft carpet; and carpet 4 is an Apprentice carpet.

92. OR RNB, f. 550, F.III.111, ll. 9–9ob, 16–17.

93. Ibid., F.III.111, l. 23ob.

94. Ibid., F.III.111, l. 11ob.

95. Ibid., F.III.113, ll. 14, 16.

96. Ibid., O.III.159, ll. 2ob–3ob.

97. Ibid., F.III.111, ll. 15ob–16.

98. Ibid., F.III.48, ll. 117–117ob.

99. *Journey to the Land of Ophir* (1784), quoted in A. Lentin, "Introduction," Prince M. M. Shcherbatov, *On the Corruption of Morals in Russia* (Cambridge, 1969), 78.

100. D. M. Griffiths, "Catherine II: The Republican Empress," *Jahrbücher für Geschichte Osteuropas*, n.s., 21, Heft 3 (1973): 335.

101. *Documents of Catherine the Great: The Correspondence with Voltaire and the "Instruction" of 1767 in the English Text of 1768*, ed. W. F. Reddaway (Cambridge, 1931), 297.

102. Ibid., 233.

103. Ibid., 225, 255.

104. See Pavel N. Miliukov, "Educational Reforms," in *Catherine the Great: A Profile*, ed. Marc Raeff (New York, 1972), 93–112; Max J. Okenfuss, "Education and Empire: School Reform in Enlightened Russia," *Jahrbücher für Geschichte Osteuropas*, n.s., 27, Heft 1 (1979): 41–68; and idem, "Popular Educational Tracts in Enlightenment Russia: A Preliminary Survey," *Canadian-American Slavic Studies* 14, no. 3 (fall 1980): 307–26. *O dolzhnostiakh cheloveka i grazhdanina, kniga k chteniiu opredelennaia v narodnykh gorodskikh uchilishchakh Rossiiskoi Imperii, izdannaia po Vysochaishemu Poveleniiu Tsarstvuiushchei Imperatritsy Ekateriny Vtoryia* (St. Petersburg, 1783). An English translation can be found in J. L. Black, *Citizens for the Fatherland: Education, Educators, and Pedagogical Ideals in Eighteenth Century Russia*, East European Monographs, 53 (New York, 1979), 209–66.

105. *O dolzhnostiakh cheloveka*, 14–67.

106. *ES*, s.v. "Universitet"; *Istoriia Moskovskogo Universiteta*, vol. 1 (Moscow, 1955), 39; R. P. Bartlett, "Culture and Enlightenment: Julius von Canitz and the Kazan' *Gimnazii* in the Eighteenth Century," *Canadian-American Slavic Studies* 14, no. 3 (fall 1980): 347.

107. *PSZ*, 1st Series, vol. 21, 1781–1783, No. 15.379, §41.

108. Pimen Arapov, comp., *Letopis' russkago teatra* (St. Petersburg, 1861), 100–101, 101 n. 1.

109. Priscilla Roosevelt, *Life on the Russian Country Estate: A Social and Cultural History* (New Haven, 1995), 145.

110. "Mysli filosofa po mode, ili sposob kazat'sia razumnym, ne imeia ni kapli razuma," in Iu. V. Stennik, *Russkaia satiricheskaia proza XVIII veka* (Leningrad, 1986), 288–95. See also "Nastavlenie molodomu suetonu, vstupaiushchemu v svet," *Vecherniaia zaria*, pt. 3 (1782): 70–73.

111. This sizable and fascinating body of literature has been largely overlooked by both literary scholars and historians of Russia. Exceptions include M. I. Demkov, *Istoriia russkoi pedagogii*, 2 vols. (St. Petersburg, 1896–1897); G. A. Kosmolinskaia, "Pamiatniki uchebno-vospitatel'noi literatury vtoroi poloviny XVIII veka v sobranii nauchnoi biblioteki MGU," in *Iz fonda redkikh knig i rukopisei nauchnoi biblioteki Moskovskogo Universiteta* (Moscow, 1987), 5–42. A useful introduction to the subject in the German context is Wolfgang Martens, *Die Botschaft der Tugend: Die Aufklärung im Spiegel der deutschen moralischen Wochenschriften* (Stuttgart, 1968).

112. [Georgii Bazhanov, trans.], *Karmannaia knizhka chestnago cheloveka, ili Nuzhnyia pravila vo vsiakom meste i vo vsiakoe vremia* (St. Petersburg, 1794); *Druzheskie sovety molodomu cheloveku, nachinaiushchemu zhit' v svete* (Moscow, 1762); *Dolzhnosti chestnago cheloveka* (St. Petersburg, 1798).

113. *Karmannaia knizhka,* 27–28; *Druzheskie sovety,* 46–47, 51. The *chestnyi chelovek* as a cultural type traces its origins back to the *honnête homme* of seventeenth-century France, although having undergone during this eastward migration a rather thorough transformation. For some interesting observations on the *chestnyi chelovek* in the early nineteenth century, see William Mills Todd III, *Fiction and Society in the Age of Pushkin: Ideology, Institutions, and Narrative* (Cambridge, Mass., 1986), 3–4, 33–34. On the French origins and subsequent history of the ideal, see Anette Höfer and Rolf Reichardt, "Honnête homme, Honnêteté, Honnête gens," in *Handbuch politisch-sozialer Grundbegriffe in Frankreich, 1680–1820,* ed. Rolf Reichardt and Eberhard Schmitt, Heft 7 (Munich, 1986), 1–67; Nannerl O. Keohane, *Philosophy and the State in France: The Renaissance to the Enlightenment* (Princeton, 1980), 231–32, 283–85.

114. Pt. 9 (August 1780), 207. The articles are in pt. 1 (September 1777), 69–83 and pt. 6 (August 1779), 293–304.

115. *SK,* vol. 4 (Moscow, 1966), nos. 150, 210, and 238. The *Companion of Lovers of the Russian Word* characterized its goals as threefold: first, to help spread knowledge and enlightenment; second, to aid the development of the Russian language; and third, to further the improvement of morals and manners.

116. For the standard view of Pnin and excerpts on his views of the passions and the fashioning of good citizens, see Vl. Orlov, "Ivan Pnin," chap. 2 in *Russkie prosvetiteli 1790–1800-kh godov* (n.p., 1950). Pnin's translations, such as *O nravouchenii, dolzhnostiakh i obiazannostiakh nravstvennykh,* and other publications can be found in Ivan Pnin, *Sochineniia,* ed. I. K. Luppol (Moscow, 1934).

117. Charles Taylor, *Sources of the Self: The Making of the Modern Identity* (Cambridge, Mass., 1989), 159–60.

118. Gerhard Oestreich, *Neostoicism and the Early Modern State,* ed. Brigitta Oestreich and H. G. Koenigsberger, trans. David McLintock (Cambridge, 1982), 7.

119. Ibid., 13, 15, 21, 33.

120. Ibid., 37–38.

121. *Istinnoe povestvovanie, ili zhizn' Gavriila Dobrynina, im samim napisannaia v Mogileve i v Vitebske, 1752–1823* (St. Petersburg, 1872), 194–98. See also I. V. Lopukhin's defense of Freemasonry before the governor-general of Moscow, Ia. A. Brius, in I. V. Lopukhin, *Zapiski senatora I. V. Lopukhina* (1859; reprint, Moscow, 1990), 25.

122. Oestreich, *Neostoicism,* 269–70. The expression comes from Marc Raeff's important study *The Well-Ordered Police State: Social and Institutional Change through Law in the Germanies and Russia, 1600–1800* (New Haven, 1983). Raeff mentions the importance of Neostoicism in eighteenth-century Russia in *Imperial Russia, 1682–1825: The Coming of Age of Modern Russia* (New York, 1971), 141–42. The place of Stoicism in the worldview of Russian Masons is noted in Ransel, *Politics of Catherinian Russia,* 55–57, 213–14; N. F. Utkina et al., *Russkaia mysl' v vek prosveshcheniia* (Moscow, 1991), 163–65. For more on Neostoicism in the intellectual world of Catherinian Russia, see Walter J. Gleason, *Moral Idealists, Bureaucracy, and Catherine the Great* (New Brunswick, 1981), 59–61.

123. Oestreich, *Neostoicism,* 7–8.

124. See Norman Fiering, *Moral Philosophy at Seventeenth-Century Harvard: A Discipline in Transition* (Chapel Hill, N.C., 1981), 3–7; J. B. Schneewind, introduction to *Moral Philosophy from Montaigne to Kant: An Anthology,* 2 vols. (Cambridge, 1990).

125. J. G. A. Pocock, "Clergy and Commerce: The Conservative Enlightenment in England," in *L'età dei lumi: Studi storici sul settecento europeo in onore di Franco Venturi* (Naples, 1985), 1:525–62; G. J. Barker-Benfield, "The Reformation of Male Manners," chap. 2 in *The Culture of Sensibility: Sex and Society in Eighteenth-Century Britain* (Chicago, 1992). Also worth consulting are Lawrence E. Klein, *Shaftesbury and the Culture of Politeness: Moral Discourse and Cultural Politics in Early Eighteenth-Century England* (Cambridge, 1994); John Mullan, "Sympathy and the Production of Society," chap. 1 in *Sentiment and Sociability: The Language of Feeling in the Eighteenth Century* (New York, 1988).

126. Barker-Benfield, *Culture of Sensibility*, 55–58.

127. Jean Starobinski, *Blessings in Disguise; or, The Morality of Evil*, trans. Arthur Goldhammer (Cambridge, Mass., 1993), 8–12. Peter France has made similar observations in "Polish, Police, *Polis*," chap. 4 in *Politeness and Its Discontents: Problems in French Classical Culture*, Cambridge Studies in French (Cambridge, 1992).

128. *Utrennii svet*, pt. 9 (June 1780): 91.

129. V. N. Zinov'ev, "Zhurnal puteshestviia V. N. Zinov'eva po Germanii, Italii, Frantsii i Anglii v 1784–1788 gg.," *RS* 23 (November 1878): 439.

130. Ibid., 213, 601–5.

131. Ibid., 207.

132. Charles François Philibert Masson, *Secret Memoirs of the Court of Petersburg* (New York, 1970), 51–52, 222.

133. Larry Wolff, *Inventing Eastern Europe: The Map of Civilization on the Mind of the Enlightenment* (Stanford, 1994), 6, 193.

134. Ibid., 22.

135. For the influential role of civility in defining national character among Russians at this time, see Hans Rogger, "Manners and Morals," chap. 2 in *National Consciousness in Eighteenth-Century Russia* (Cambridge, Mass., 1960).

136. Shcherbatov, *On the Corruption of Morals in Russia*, 134, 136, 142.

137. France, "Polish, Police, *Polis*," 64–66; Starobinski, *Blessings in Disguise*, 10–11. On the Russian context, see L. A. Chernaia, "Kontseptsiia lichnosti v russkoi literature vtoroi poloviny XVII-pervoi poloviny XVIII v.," in *Razvitie Barokko i zarozhdenie Klassitsizma v Rossii XVII-nachala XVIII v.*, ed. A. N. Robinson (Moscow, 1989), 220–32.

2. THE RUSSIAN PUBLIC; OR, CIVIL SOCIETY IN THE EIGHTEENTH CENTURY

1. N. I. Novikov, *Satiricheskie zhurnaly N. I. Novikova*, ed. P. N. Berkov (Moscow-Leningrad, 1951), 68.

2. W. Gareth Jones, "The Polemics of the 1769 Journals: A Reappraisal," *Canadian-American Slavic Studies* 16, nos. 3–4 (fall–winter 1982): 432–43.

3. The most recent examination of the Novikov affair is K. A. Papmehl, "The Empress and 'Un Fanatique': A Review of the Circumstances Leading to the Government Action against Novikov in 1792," *Slavonic and East European Review* 68, no. 4 (October 1990): 665–91. See also W. Gareth Jones, *Nikolay Novikov, Enlightener of Russia* (Cambridge, 1984), 206–11.

4. It is interesting to note that this perception is generally shared by those at both ends of the political spectrum. At one end, see, for example, Richard Pipes,

Russia under the Old Regime (New York, 1974). At the other, see Geoff Eley, "Nations, Publics, and Political Cultures: Placing Habermas in the Nineteenth Century," in *Habermas and the Public Sphere,* ed. Craig Calhoun (Cambridge, Mass., 1992), 325.

5. Along these lines it behooves one to heed, if not necessarily to follow, Gilbert Rozman's admonition to lay to rest the "myth of Russian backwardness," *Urban Networks in Russia, 1750–1800, and Premodern Periodization* (Princeton, 1976), 279. On the generally unrecognized vibrancy of Russia's urban and commercial life, see Elise Kimerling Wirtschafter, *Social Identity in Imperial Russia* (DeKalb, Ill., 1997), 74–75, 81–82. In the same vein, "backwardness" is always a matter of perspective. See, for example, a discussion of France's public that stresses the danger of falling back on past historiographical interpretations focusing on the supposed underdevelopment of France vis-à-vis England. Daniel Gordon, David A. Bell, and Sarah Maza, "The Public Sphere in the Eighteenth Century," *French Historical Studies* 17, no. 4 (fall 1992): 882–956.

6. For the eighteenth century see Malcolm Burgess, "Russian Public Theater Audiences of the Eighteenth and Early Nineteenth Centuries," *Slavonic and East European Review* 37, no. 88 (December 1958): 160–83; Gary Marker, *Publishing, Printing, and the Origins of Intellectual Life in Russia, 1700–1800* (Princeton, 1985); Marc Raeff, "Transfiguration and Modernization: The Paradoxes of Social Disciplining, Paedagogical Leadership, and the Enlightenment in Eighteenth-Century Russia," in *Alteuropa—Ancien Régime—Frühe Neuzeit: Probleme und Methoden der Forschung,* ed. Hans Erich Bödeker and Ernst Hinrichs (Stuttgart, 1991), 99–115. On the nineteenth and twentieth centuries see Jeffrey Brooks, *When Russia Learned to Read: Literacy and Popular Literature, 1861–1917* (Princeton, 1985); Edith W. Clowes, Samuel D. Kassow, and James L. West, eds., *Between Tsar and People: Educated Society and the Quest for Identity in Late Imperial Russia* (Princeton, 1991); Manfred Hagen, *Die Entfaltung politischer Öffentlichkeit in Russland, 1906–1914* (Wiesbaden, 1982); Louise McReynolds, *The News under Russia's Old Regime: The Development of a Mass Circulation Press* (Princeton, 1991); Nicholas V. Riasanovsky, *A Parting of Ways: Government and the Educated Public in Russia, 1801–1855* (Oxford, 1976); Grigory Sternin, "Public and Artist in Russia at the Turn of the Century," in *Tekstura: Russian Essays on Visual Culture,* ed. and trans. Alla Efimova and Lev Manovich (Chicago, 1993), 89–114; William Mills Todd III, *Fiction and Society in the Age of Pushkin: Ideology, Institutions, and Narrative* (Cambridge, Mass., 1986).

7. Riasanovsky, *Parting of Ways,* 22.

8. Jürgen Habermas, *The Structural Transformation of the Public Sphere: An Inquiry into a Category of Bourgeois Society,* trans. Thomas Burger, with the assistance of Frederick Lawrence (Cambridge, Mass., 1989). Anthony J. La Vopa provides a useful overview of the influence Habermas, as well as Reinhart Koselleck, has had on recent scholarship in "Conceiving a Public: Ideas and Society in Eighteenth-Century Europe," *Journal of Modern History* 64 (March 1992): 79–116. The broader implications of Habermas's work are addressed in Calhoun, ed., *Habermas and the Public Sphere.* See also Bruce Robbins, ed., *The Phantom Public Sphere* (Minneapolis, 1993) for a recent critique of the concept of the public as elaborated by Habermas and others.

9. Habermas, *Structural Transformation,* 17–18.

10. N. A. Smirnov, "Zapadnoe vliianie na russkii iazyk v petrovskuiu epokhu," *Sbornik otdeleniia russkago iazyka i slovesnosti Imperatorskoi Akademii nauk* 88,

no. 2 (St. Petersburg, 1910), 248–49.

11. *PSZ*, vol. 6, 1720–1722, No. 3534.

12. Smirnov, "Zapadnoe vliianie," 248.

13. Marc Raeff, *The Well-Ordered Police State: Social and Institutional Change through Law in the Germanies and Russia, 1600–1800* (New Haven, 1983). See also the excellent study by John P. LeDonne, *Absolutism and Ruling Class: The Formation of the Russian Political Order, 1700–1825* (New York, 1991).

14. Habermas, *Structural Transformation*, 18–22.

15. Ivan Nordstet, *Rossiiskii, s nemetskim i frantsuzskim perevodami, slovar', sochinennyi nadvornym sovetnikom Ivanom Nordstetom*, pt. 2 (St. Petersburg, 1782), s.v. "Publika."

16. *Slovar' Akademii Rossiiskoi*, pt. 4 (St. Petersburg, 1794), s.v. "Obshchestvo."

17. Nordstet, *Rossiiskii, s nemetskim*, pt. 2, s.v. "Obshchestvo."

18. Lucian Hölscher, *Öffentlichkeit und Geheimnis: Eine begriffsgeschichtliche Untersuchung zur Entstehung der Öffentlichkeit in der frühen Neuzeit* (Stuttgart, 1979), 83–89; idem, "Öffentlichkeit," in *Geschichtliche Grundbegriffe: Historisches Lexikon zur politisch-sozialen Sprache in Deutschland*, ed. Otto Brunner, Werner Conze, and Reinhart Koselleck, Band 4 (Stuttgart, 1978), 430–31.

19. Hölscher, *Öffentlichkeit und Geheimnis*, 83, 88–89; idem, "Öffentlichkeit," 431–35. Habermas observes that it was in the new "centers of criticism" (coffeehouses, salons, and so on) that there began to emerge "between aristocratic society and bourgeois intellectuals, a certain parity of the educated" (*Structural Transformation*, 32). For the most thorough recent discussion of German civil society in the eighteenth century that builds upon and modifies Habermas's formulation, see Isabel V. Hull, "The Practitioners of Civil Society," chap. 5 in *Sexuality, State, and Civil Society in Germany, 1700–1815* (Ithaca, N.Y., 1996).

20. *Sankt Peterburgskii Merkurii*, pt. 2 (1793): 222.

21. *Sanktpeterburgskiia vedomosti*, 19 May 1780, no. 40, *Pribavlenie*, 529.

22. Vladimir Dal', *Tolkovyi slovar' zhivago velikorusskago iazyka*, vol. 3 (Moscow, 1882), s.v. "Publika."

23. Abbott Gleason, "The Terms of Russian Social History," in Clowes et al., *Between Tsar and People*, 18–19; LeDonne, *Absolutism and Ruling Class*, viii. Anthony Graham Netting defines *obshchestvo* in the reigns of Catherine II and Alexander I as referring simply to "the high aristocratic circles around the court." "Russian Liberalism: The Years of Promise, 1842–1855" (Ph.D. diss., Columbia University, 1967), 20.

24. Raeff, "Transfiguration and Modernization," 109. The internal quotes are Raeff's.

25. Wirtschafter, *Social Identity*, 36, 59.

26. LeDonne, *Absolutism and Ruling Class*, 5–16, 21, 158; LeDonne, *Ruling Russia: Politics and Administration in the Age of Absolutism, 1762–1796* (Princeton, 1984), 18–19.

27. Habermas, *Structural Transformation*, 22–23. See also Robert Wuthnow, *Communities of Discourse: Ideology and Social Structure in the Reformation, the Enlightenment, and European Socialism* (Cambridge, Mass., 1989), 171–79, 201–3, 310–13. Wuthnow, who highlights the state's role in creating and buttressing the public sphere, defines the public as formed of "public officials, administrators, parliamentary representatives, courtiers, lawyers, professionals, military officers, men and women of

leisure, [and] university faculty."

28. H. P. Sumarokov, *Polnoe sobranie vsekh sochinenii,* ed. N. Novikov, 2nd ed., pt. 4 (Moscow, 1787), 61–62.

29. Ibid., 4:63. A comparable image of the public was that of a mindless bunch of imitators and followers, greedily chasing after the latest novelty. See, for example, Catherine's brief greeting to "Mrs. Public" in *All Sorts of Things* from 1769, reprinted in *Sochineniia Imperatritsy Ekateriny II,* ed. A. N. Pypin, vol. 5 (St. Petersburg, 1903), 281.

30. According to one study, although Russia remained overwhelmingly rural, the number of Russian urban dwellers approximately doubled between 1700 and 1800. J. Michael Hittle, *The Service City: State and Townsmen in Russia, 1600–1800* (Cambridge, Mass., 1979), 178. Gilbert Rozman has observed that by 1800 the "percentage of its [Russia's] population living in cities was well above the world average" and that Russia had "joined the ranks of a small number of countries as relatively urbanized, advanced premodern societies" (*Urban Networks,* 8, 277).

31. Ulrich Im Hof, *Das gesellige Jahrhundert: Gesellschaft und Gesellschaften im Zeitalter der Aufklärung* (Munich, 1982).

32. Marker, *Publishing,* 19.

33. Ibid., 76, 89–90, 105, 138.

34. Ibid., 71, 105.

35. Ibid., 138.

36. N. D. Chechulin, *Russkoe provintsial'noe obshchestvo vo vtoroi polovine XVIII veka* (St. Petersburg, 1889), 76.

37. Marker, *Publishing,* 150–51, 178, 180–83, 233–36. See also I. F. Martynov, "Kniga v russkoi provintsii 1760–1790-kh gg.: Zarozhdenie provintsial'noi knizhnoi torgovli," in *Kniga v Rossii do serediny XIX v.,* ed. A. A. Sidorov and S. P. Luppov (Leningrad, 1978), 109–25. *Zerkalo sveta* pt. 1, no. 1 (9 February 1786). According to this issue, subscriptions were sent to over thirty provincial centers.

38. Marker, *Publishing,* 24, 230.

39. Ibid., 202.

40. Ibid., 118–20, 201–6. Marker even refers to the development of an "early 'Grub Street' literature in Russia" around this time. For similar sentiments among the English during the same period, see John Brewer, *The Pleasures of the Imagination: English Culture in the Eighteenth Century* (New York, 1997), 192–94.

41. *Vzgliad na moiu zhizn'* (Moscow, 1866), 34. Quoted in Henry M. Nebel, Jr., *N. M. Karamzin: A Russian Sentimentalist* (The Hague, 1967), 17.

42. Jones, *Nikolay Novikov,* 31, 48.

43. *PSZ,* vol. 5, 1713–1719, No. 3246. Unfortunately, the assemblies have yet to find their modern historian. For a brief discussion, see L. N. Semenova, *Ocherki istorii byta i kul'turnoi zhizni Rossii (pervaia polovina XVIII v.)* (Leningrad, 1982), 199–206. Descriptions of the assemblies by two participants can be found in the memoirs of Friedrich Wilhelm von Bergholz, *Dnevnik kamer"-iunkera F. V. Berkhgol'tsa, 1721–1725* (Moscow, 1902–1903); Christian Weber, "Zapiski Vebera o Petre Velikom i ego preobrazovaniiakh," *RA,* bk. 1 (1872): 1057–168; bk. 2 (1872): 1334–457, 1613–704.

44. *PSZ,* vol. 5, 1713–1719, No. 3246.

45. In addition to Golitsyn, Artamon Matveev (a favorite of Tsar Alexis Mikhailovich) and his English wife instituted a sort of salon in their Moscow home attended by members of both sexes. Liubov' Gurevich, *Istoriia russkogo teatral'nogo*

byta, vol. 1 (Moscow-Leningrad, 1939), 10.

46. The ukase contained seven points that set out the specific workings of the gatherings and were to be strictly followed "until they became customary."

47. Semenova, *Ocherki istorii,* 161–74.

48. Ibid., 200–201.

49. Nancy Shields Kollmann, "The Seclusion of Elite Muscovite Women," *Russian History* 10, pt. 2 (1983): 170, 174. In an important recent study, Elise Kimerling Wirtschafter discusses the integral role of women in eighteenth-century *obshchestvo. Social Identity,* 13–19.

50. Emile Haumant, *La culture française en Russie (1700–1900),* 2nd ed. (Paris, 1913), 83.

51. M. I. Pyliaev, *Staryi Peterburg* (1889; reprint, Leningrad, 1990), 410.

52. Ibid., 168–73.

53. L. I. Kulakova, "M. M. Kheraskov," in *Istoriia russkoi literatury,* vol. 4 (Moscow-Leningrad, 1947), 321–22; A. V. Zapadov, "Tvorchestvo Kheraskova," in *Izbrannye proizvedeniia,* by M. M. Kheraskov (Leningrad, 1961), 9; P. N. Berkov, *Istoriia russkoi zhurnalistiki XVIII veka* (Moscow-Leningrad, 1952), 291–98.

54. Pyliaev, *Staryi Peterburg,* 293–304 (Vigel quoted p. 304).

55. Semenova, *Ocherki istorii,* 205–6. The *St. Petersburg News* for 17 April 1780, no. 31, *Pribavlenie,* 419, carried an advertisement for an upcoming masquerade "in the home of the Free Economic Society" on the 22nd of the month. The price of admission was set at one ruble per couple. According to A. V. Koval'chuk, a specialist in eighteenth-century Russian economic history at the Russian Academy of Science's Institute of History in Moscow, during the reign of Catherine II the state spent approximately nineteen rubles to house, feed, and clothe a soldier for a year. This gives some indication of the relative cost to attend such balls.

56. Pyliaev, *Staryi Peterburg,* 230.

57. Ibid., 143–45; *Sanktpeterburgskiia vedomosti,* 25 February 1780, no. 16, *Pribavlenie,* 201; see also 13 March, no. 21, 258–59.

58. *Sanktpeterburgskiia vedomosti,* 26 May 1780, no. 42, *Pribavlenie,* 565.

59. *Prazdnoe vremia v pol'zu upotreblennoe* 1 (1759): 365–66.

60. Pyliaev, *Staryi Peterburg,* 432–35.

61. Gurevich, *Istoriia russkogo teatral'nogo byta,* 1:7–26; A. S. Eleonskaia, "Tvorcheskie vzaimosviazi shkol'nogo i pridvornogo teatrov v Rossii," in *P'esy stolichnykh i provintsial'nykh teatrov pervoi poloviny XVIII v.,* ed. O. A. Derzhavina et al. (Moscow, 1975), 7–14. Tsarina Praskov'ia Fedorovna operated a theater similar to that of Natal'ia Alekseevna at Izmailov between 1713 and 1723.

62. Gurevich, *Istoriia russkogo teatral'nogo byta,* 1:26–30; Eleonskaia, "Tvorcheskie vzaimosviazi," 7–11.

63. B. N. Aseev, *Russkii dramaticheskii teatr ot ego istokov do kontsa XVIII veka* (Moscow, 1977), 231–32.

64. By an ukase of 15 June 1751, Elizabeth permitted into the theater Russian and foreign merchants and their wives "as long as they are not foully attired." Quoted in V. Vsevolodskii-Gerngross, *Russkii teatr ot istokov do serediny XVIII veka* (Moscow, 1957), 192.

65. Burgess, "Russian Public Theater," 162–63; Vsevolodskii-Gerngross, *Russkii teatr,* 192, 221–28.

66. Burgess, "Russian Public Theater," 182; Vsevolodskii-Gerngross, *Istoriia russkogo dramaticheskogo teatra,* vol. 1, *Ot istokov do kontsa XVIII veka* (Moscow, 1977), 265–67, 277–78.

67. Burgess, "Russian Public Theater," 162; L. N. Maikov, *Ocherki iz istorii russkoi literatury XVII i XVIII stoletii* (St. Petersburg, 1889), 310–13; B. V. Varnecke, *History of the Russian Theatre* (New York, 1951), 110.

68. Burgess, "Russian Public Theater," 178–79; Maikov, *Ocherki,* 315.

69. See the dramatist Vladimir Lukin's colorful and decidedly negative description of some of the characters he encountered at one performance. Burgess, "Russian Public Theater," 177–78.

70. Ibid., 175–78; Vsevolodskii-Gerngross, *Istoriia,* 1:248, 266–67.

71. Maikov, *Ocherki,* 315; Vsevolodskii-Gerngross, *Istoriia,* 1:267–69.

72. *Istoriia Akademii nauk SSSR,* vol. 1 (Moscow-Leningrad, 1958), 52, 172–74, 330. One of these announcements from 1761 is reproduced on p. 173.

73. T. V. Staniukovich, *Kunstkamera peterburgskoi Akademii nauk* (Moscow-Leningrad, 1953), 58, 68–73, 162–64, 198–99. The internal quotes are drawn from *Materialy dlia istorii Akademii nauk,* 10 vols. (St. Petersburg, 1885–1900).

74. *Istoriia Biblioteki Akademii nauk SSSR, 1714–1964,* pt. 1 (Moscow-Leningrad, 1964), 44–45, 85–86, 117, 157.

75. A. A. Zaitseva, "'Kabinety dlia chteniia' v Sankt-Peterburge kontsa XVIII–nachala XIX veka," in *Russkie biblioteki i chastnye knizhnye sobraniia XVI–XIX vekov,* ed. D. V. Ter-Avanesian (Leningrad, 1979), 30; K. I. Abramov, *Istoriia bibliotechnogo dela v SSSR* (Moscow, 1970), 43. There were also libraries at the Academy of Arts, the Academic high schools, and the School of Mines. *Istoriia Biblioteki Akademii nauk,* 128.

76. Zaitseva, "'Kabinety dlia chteniia,'" 30–37, 43–45.

77. For two entertaining descriptions of Moscow society in the eighteenth century, see I. Zabelin, "Khronika obshchestvennoi zhizni v Moskve s poloviny XVIII stoletiia," in *Opyty izucheniia russkikh drevnostei i istorii,* pt. 2 (Moscow, 1873): 351–506; idem, "Iz khroniki obshchestvennoi zhizni v Moskve v XVIII stoletii," in *Sbornik obshchestva liubitelei rossiiskoi slovesnosti na 1891 god* (Moscow, 1891): 557–82.

78. Perhaps not surprisingly, Ianovskii's actions did not fail to arouse the ire of some of his colleagues. See the pointed remarks from 1731 of Sil'vestr Kholmskii, the Metropolitan of Kiev, in Semenova, *Ocherki istorii,* 204–5.

79. L. Starikova, *Teatral'naia zhizn' starinnoi Moskvy: Epokha, byt, nravy* (Moscow, 1988), 258–63.

80. Burgess, "Russian Public Theater," 168.

81. M. I. Pyliaev, *Staraia Moskva: Rasskazy iz byloi zhizni pervoprestol'noi stolitsy* (1891; reprint, Moscow, 1990), 40–41.

82. A. Lentin, introduction to *On the Corruption of Morals in Russia,* by M. M. Shcherbatov (Cambridge, 1969), 40; M. N. Longinov, *Novikov i moskovskie martinisty* (Moscow, 1867), 125.

83. *Moskovskiia vedomosti,* 31 May 1756, no. 11; and 18 June 1756, no. 16.

84. Quoted in Starikova, *Teatral'naia zhizn',* 251.

85. Zabelin, "Khronika obshchestvennoi zhizni v Moskve," 405.

86. W. Tooke, *History of Russia: From the Foundation of the Monarchy by Riurik to the Accession of Catherine the Second,* vol. 2 (London, 1800), 430.

87. Pyliaev, *Staraia Moskva*, 125–30.

88. Starikova, *Teatral'naia zhizn'*, 252.

89. On Maddox, see the contradictory statements in Pyliaev, *Staraia Moskva*, 391 n. 23; Tooke, *History of Russia*, 2:404; Anthony Cross, *By the Banks of the Neva: Chapters in the Lives and Careers of the British in Eighteenth-Century Russia* (Cambridge, 1997).

90. Pyliaev, *Staraia Moskva*, 326–27; Tooke, *History of Russia*, 2:427–28.

91. Starikova, *Teatral'naia zhizn'*, 266–68; Vsevoldskii-Gerngross, *Istoriia*, 1:278–80. On the early history of Moscow theater, see Aseev, *Russkii dramaticheskii teatr*, 118–34; Gurevich, *Istoriia russkogo teatral'nogo byta*, 1:7–30.

92. Burgess, "Russian Public Theater," 182.

93. Tooke, *History of Russia*, 2:404–6. Subscriptions to the loges were advertised in the *Moscow News*. See Pyliaev, *Staraia Moskva*, 103. Premieres at the Petrovskii, such as that of *The Marriage of Figaro* on 15 January 1787, were major social events attracting such large crowds that, according to one theatergoer, "there wasn't even room for an apple to fall *[iabloku negde bylo upast']*." See Starikova, *Teatral'naia zhizn'*, 268.

94. Quoted in Burgess, "Russian Public Theater," 170–71. See also Priscilla Roosevelt, *Life on the Russian Country Estate: A Social and Cultural History* (New Haven, 1995), 135.

95. Brewer, *Pleasures of the Imagination*, 396–400.

96. Tooke, *History of Russia*, 2:404–6, 451–52; Pyliaev, *Staraia Moskva*, 103–4; Starikova, *Teatral'naia zhizn'*, 251–52, 263, 266–78.

97. S. P. Shevyrev, *Istoriia Imperatorskago Moskovskogo Universiteta* (Moscow, 1855), 66, 568–69, quoted in J. Thomas Sanders, "The Third Opponent: Dissertation Defenses and the Public Profile of Academic Life in Late Imperial Russia," *Jahrbücher für Geschichte Osteuropas*, n.s., 41, Heft 2 (1993): 243–45.

98. Shevyrev, *Istoriia*, 66, quoted in Sanders, "The Third Opponent," 245.

99. *Istoriia Moskovskogo Universiteta*, vol. 1 (Moscow, 1955), 60.

100. Sanders, "The Third Opponent," 245–46; A. A. Kizevetter, "Moskovskii Universitet (istoricheskii ocherk)," in *Moskovskii Universitet, 1755–1930: Iubileinyi sbornik*, ed. V. B. El'iashevich, A. A. Kizevetter, and M. M. Novikov (Paris, 1930), 43–44, 60–61.

101. Kizevetter, "Moskovskii Universitet," 15, 47–52; *Istoriia Moskovskogo Universiteta*, 1:60.

102. *Moskovskiia vedomosti*, 2 July 1756, no. 20, *Pribavlenie*, 3; *Istoriia Moskovskogo Universiteta*, 1:65.

103. Abramov, *Istoriia*, 42–43; Marker, *Publishing*, 172. The publisher V. V. Kiprianov established Russia's first public library in Moscow in 1727. See Abramov, *Istoriia*, 33–34.

104. John LeDonne writes that in the years following 1762 "a new field of action was opening up in the provinces and districts: the close management of the peasantry and the constitution of a civil society within the ruling class." *Absolutism and Ruling Class*, 304.

105. Isabel de Madariaga, *Russia in the Age of Catherine the Great* (New Haven, 1981), 290.

106. LeDonne, *Absolutism and Ruling Class*, 26–27; A. Romanovich-Slavatinskii,

Dvorianstvo v Rossii ot nachala XVIII veka do otmeny krepostnogo prava, 2nd ed. (Kiev, 1912), 437–54; Roosevelt, *Life on the Russian Country Estate*, 200.

107. On this transformation in the life of the Russian provinces, see Chechulin, *Russkoe provintsial'noe obshchestvo*, 72–76; Vsevolodskii-Gerngross, *Istoriia*, 1:272–75. The rhythms of Smolensk's local society in the Catherinian age, from its informal salons and dinner parties to its Gentry Club and amateur theater productions, are depicted in a little-known novel by [F. A. Ettinger] F.v.E., *Bashnia Veselukha, ili Smolensk i zhiteli ego shestdesiat' let nazad* (1845; reprint, Smolensk, 1992). On this novel and life in Smolensk, see Richard Stites, "Old Smolensk: Two Sketches and Notes for Research," unpublished paper presented at the workshop "Visions, Institutions, and Experiences of Imperial Russia," Washington, D.C., September 1993.

108. Vsevolodskii-Gerngross, *Istoriia*, 1:273–74; Varnecke, *History*, 96–108; R. M. Lazarchuk, "Iz istorii provintsial'nogo teatra (Vologodskii publichnyi teatr kontsa XVIII–nachala XIX v.)," *XVIII vek* 18 (St. Petersburg, 1993), 156–71; idem, "Iz istorii provintsial'nogo teatra (teatral'naia zhizn' Vologdy 1780-kh gg.)," *XVIII vek* 15 (Leningrad, 1986), 52–69.

109. Vsevolodskii-Gerngross, *Russkii teatr*, 218–19, 240–41.

110. Gregory L. Freeze, *The Russian Levites: Parish Clergy in the Eighteenth Century* (Cambridge, Mass., 1977), 102. See, for example, the description of the "public act" staged in 1759 at the Kazan gymnasium and the public celebration from a year later that included, among other amusements, Mercury descending via a thin rope from the peak of an artificial mountain. A. I. Artem'ev, "Kazanskiia gimnazii v XVIII stoletii," *Zhurnal Ministerstva narodnago prosveshcheniia* 173 (May 1874): 68–69, 74–76.

111. Abramov, *Istoria*, 42; *Istoriia Biblioteki Akademii nauk*, 128; Marker, *Publishing*, 172.

112. Nicholas Hans, "The Moscow School of Mathematics and Navigation (1701)," *Slavonic and East European Review* 29, no. 73 (1951): 532–36.

113. Pyliaev, *Staryi Peterburg*, 222–24. The novelty of clubs is humorously conveyed by an early play of D. I. Fonvizin, in which a dim-witted Ulita confuses the new "kloby" (clubs) with "klopy" (bedbugs). *Dramatic Works of D. I. Fonvizin*, ed. and trans. Marvin Kantor (Frankfurt a. M., 1974), 26.

114. Pyliaev, *Staryi Peterburg*, 226; *ES*, s.v. "Klub."

115. Longinov, *Novikov*, 168 n. 38, 304. Perhaps Melissino's society drew its inspiration from the Most Ancient and Puissant Order of the Beggar's Benison and Merryland, a Scottish secret society devoted to group masturbation, voyeurism, and the reading of pornographic literature. British expatriates may have founded a St. Petersburg branch of the order in the reign of Catherine II. Cross, *By the Banks of the Neva*, 34–36.

116. Pyliaev, *Staryi Peterburg*, 228–29.

117. *Stoletie S. Peterburgskago angliiskago sobraniia, 1770–1870* (St. Petersburg, 1870), 1–2.

118. For a list of all the assembly's members between 1770 and 1870, see the appendix in *Stoletie*. Although its statutes make no specific reference to excluding women, the fact that no women were ever members rather convincingly shows their de facto exclusion.

119. *Stoletie*, 3–5, 8, 12; Pyliaev, *Staryi Peterburg*, 226.

120. *Stoletie*, 13–14, 25–26.

121. Ibid., 3–4.

122. Ibid., 8.

123. A. F. Malinovskii, *Obozrenie Moskvy*, compiled by S. R. Dolgova (Moscow, 1992), 161; *ES*, s.v. "Klub." According to Anthony Cross, the Moscow English Club was established in 1782. *By the Banks of the Neva*, 40–41.

124. The sources are in disagreement over the Noble Assembly's founders, some listing Prince A. B. Golitsyn in place of A. B. Dolgorukii. Pyliaev, *Staraia Moskva*, 52; Malinovskii, *Obozrenie*, 159–60; *ES*, s.v. "Klub"; Tooke, *History of Russia*, 2:448–51.

125. P. Pekarskii, *Dopolneniia k istorii masonstva v Rossii XVIII stoletiia* (St. Petersburg, 1869), 130; *Istoriia russkoi muzyki*, vol. 3 (Moscow, 1985), 404.

126. Longinov, *Novikov*, 304, 304 n. 23; A. M. Skabichevskii, *Ocherki istorii russkoi tsenzury (1700–1863 g.)* (St. Petersburg, 1892), 49.

127. N. B., "Znachenie fran-masonstva [*sic*] dlia flota," *More* 11–12 (March 1907): 315.

128. *Zakony ustanovlennago v 1792 goda sentiabria v 1 den' v gubernskom gorode Revele Kluba Soglasiia* (St. Petersburg, 1792). This booklet contains the fifty-three rules and regulations of the Harmony Club, made up of men from the "Military, civil, noble, scholarly and merchant estates." Just as the club claimed to have only men of high moral qualities as its members, only women of "honorable behavior" were permitted to attend its open social functions. For the passages cited above, see pp. 3–5, 15, 23–24.

129. On the history of the Free Economic Society, see Erich Donnert, "Anfänge der Petersburger Freien Ökonomischen Gesellschaft," chap. 3 in *Politische Ideologie der russischen Gesellschaft zu Beginn der Regierungszeit Katharinas der II* (Berlin, 1976); A. I. Khodnev, *Istoriia imperatorskago Vol'nago Ekonomicheskago Obshchestva s 1765 do 1865 goda* (St. Petersburg, 1865); Abramov, *Istoriia*, 43. The quotes, from the Society's plan and constitution, are from Donnert, p. 140.

130. On the societies initiated by Catherine and Novikov, see Jones, *Nikolay Novikov*, 82–95; I. E. Barenbaum, *Istoriia knigi*, 2nd ed. (Moscow, 1984), 57. Information on the other societies is contained in *ES*, s.v. "Sankt-Peterburgskii Vestnik"; N. Chulkov, "F. V. Krechetov—zabytyi radikal'nyi publitsist XVIII veka," in *Literaturnoe nasledstvo*, vols. 9–10 (Moscow, 1933), 456–58; V. P. Semennikov, "Literaturno-obshchestvennyi krug Radishcheva," in *Radishchev: Materialy i issledovaniia*, ed. A. S. Orlov, (Moscow-Leningrad, 1936), 212–89. A detailed but never realized plan by I. F. Bogdanovich from the early 1790s for a Society of Russian Writers was published in *Bibliograficheskiia zapiski* 3, no. 7 (1861): 194–99.

131. Berkov, *Istoriia russkoi zhurnalistiki*, 365; G. V. Vernadskii, *Russkoe masonstvo v tsarstvovanie Ekateriny II* (Petrograd, 1917), 207–8; *Biograficheskii slovar' professorov i prepodavatelei imperatorskago Moskovskogo universiteta* (Moscow, 1855), 588–89. Other student circles include the Assembly of Pupils of Moscow University's Noble Boarding School founded in 1799. See N. V. Sushkov, *Vospominaniia o Moskovskom universitetskom blagorodnom pansione* (Moscow, 1848), 12–13, 64–73.

132. *Druzheskoe Uchenoe Obshchestvo s dostodolzhnym vysokopochitaniem priglashaet sim i prosit imeniteishikh liubitelei nauk i pokrovitelei uchenosti udostoit svoim prisutstviem torzhestvennoe ego otkrytie, imeiushchee byt' v dome ego vysokorodiia, Petra Alekseevicha Tatishcheva, noiabria 6 dnia, 1782 goda* (Moscow, 1782), 7.

This rare booklet is reprinted in Longinov, *Novikov,* 04–09.

133. On the Friendly Learned Society, see B. I. Krasnobaev, "Eine Gesellschaft Gelehrter Freunde am Ende des 18. Jahrhunderts: 'Druzheskoe uchenoe obshchestvo,'" in *Beförderer der Aufklärung in Mittel- und Osteuropa: Freimaurer, Gesellschaften, Clubs,* ed. Éva H. Balázs, Ludwig Hammermayer, Hans Wagner, and Jerzy Wojtowicz (Berlin, 1979), 257–70.

134. *ES,* s.v. "Melissino, Ivan Ivanovich"; *Istoriia Moskovskogo Universiteta,* 1:65–66.

135. S. A. Tuchkov, *Zapiski* in *Zolotoi vek Ekateriny Velikoi: Vospominaniia,* ed. V. M. Bokova and N. I. Tsimbaev (Moscow, 1996), 184.

136. On Schwarz's difficulties with his academic colleagues, see Kizevetter, "Moskovskii Universitet," 47–51.

137. Ibid., 51–52; *Istoriia Moskovskogo Universiteta,* 1:65–66. For details on the Assembly of Lovers of Russian Learning, see its statutes in RGADA, f. 17, e./x. 49, ll. 1–10ob.

138. Marker, *Publishing,* 146–47; E. Shmurlo, *Mitropolit Evgenii kak uchenyi. Rannie gody zhizni, 1767–1804* (St. Petersburg, 1888), 172–81. Other towns that hosted similar literary circles were Iaroslavl', Tobolsk, and Nikolaev. See Marker, *Publishing,* 143–49. It is interesting to note that the adoption of special names was common to sodalities in other parts of Europe. On such a practice among the members of Scotland's Easy Club, see Marvin B. Becker, *The Emergence of Civil Society in the Eighteenth Century: A Privileged Moment in the History of England, Scotland, and France* (Bloomington, Ind., 1994), 70.

139. *ES,* s.v. "Fomin, A. I."; L. A. Fedorovskaia, "'Azbuk musikiiskogo peniia' iz knig Aleksandra Fomina," in *Kniga v Rossii XVI–seredina XIX v.,* ed. A. A. Zaitseva (Leningrad, 1987), 180–81; A. D. Stepanskii, *Istoriia obshchestvennykh organizatsii dorevoliutsionnoi Rossii* (Moscow, 1979), 7. One more little known provincial assembly was the Society of the Friends of Science and Usefulness established with the support of Moscow's Friendly Learned Society in the town of Orlov. A. I. Serkov, "Rossiiskoe masonstvo: Entsiklopedicheskii slovar'," vol. 1, Masony v Rossii (1731–1799)," unpublished manuscript, Moscow, n.d., 108.

140. Pyliaev, *Staryi Peterburg,* 226–28, 286.

141. Ibid., 436.

142. Grot, the grandfather of the famous academician, had earlier served as secretary to general N. A. Korf in Königsberg before coming with him to St. Petersburg sometime before 1762. He is perhaps best known for his published sermons and works on the history of non-Orthodox religions in the Russian empire. See Ia. K. Grot, "Zametka o pastore Grota," *Sbornik otdeleniia russkago iazyka i slovesnosti Imperatorskoi Akademii nauk,* vol. 5 (1868): 289–92.

143. J. C. Grot, *Uchrezhdenie osnovannago v Sanktpeterburge na smertnye sluchai obshchestva* (St. Petersburg, 1780), 1–2. Italics in the original. This text contains the society's statutes, official goals, and list of members. The information on the society's membership is drawn from the introduction and the "List of Past and Current Members." First appearing in German in 1775, the book went through two subsequent editions in Russian and French as well as German.

144. Grot, "List of Past and Current Members," *Uchrezhdenie,* 3–18.

145. Ibid., 6.

146. Ibid., 15–40.

147. See, for example, *Sanktpeterburgskiia vedomosti*, 8 September 1780, no. 72, *Pribavlenie*, 913–14; 22 September 1780, no. 76, *Pribavlenie*, 951.

148. Marquis de Custine, *Empire of the Czar: A Journey through Eternal Russia* (New York, 1989), 269.

149. Irena Grudzinska Gross, *The Scar of Revolution: Custine, Tocqueville, and the Romantic Imagination* (Berkeley–Los Angeles, 1991), 20, 41, 48, 58, 92–93.

150. See, for example, Marvin Becker's *Emergence of Civil Society in the Eighteenth Century* that, despite its title, examines the intellectual and ideological transformations that signaled the birth of European civil society.

151. Ibid., 69; Richard van Dülmen, *Die Gesellschaft der Aufklärer: Zur bürgerlichen Emanzipation und aufklärerischen Kultur in Deutschland* (Frankfurt a. M., 1986), 150–52; Daniel Roche, "Literarische und geheime Gesellschaftsbildung im vorrevolutionären Frankreich: Akademien und Logen," in *Lesegesellschaften und bürgerlichen Emanzipation: Ein europäischer Vergleich*, ed. Otto Dann (Munich, 1981), 182.

152. Adam B. Seligman, *The Idea of Civil Society* (New York, 1992), 187.

153. LeDonne, *Absolutism and Ruling Class*, ix; David L. Ransel, *The Politics of Catherinian Russia: The Panin Party* (New Haven, 1975), 1; Roosevelt, *Life on the Russian Country Estate*, 135; Wirtschafter, *Social Identity*, 42, 169.

154. *Social Identity*, 97–99.

155. *Idea of Civil Society*, 3.

156. See, for example, the essays in Clowes, Kassow, and West, eds., *Between Tsar and People*. For a useful discussion of the history of the idea of civil society, see John Keane, "Despotism and Democracy: The Origins and Development of the Distinction Between Civil Society and the State, 1750–1850," in *Civil Society and the State: New European Perspectives*, ed. John Keane (London, 1988). Also noteworthy is Seligman, *Idea of Civil Society*.

157. On one side of the debate over the subject of women and the public sphere in Europe, see Joan B. Landes, *Women and the Public Sphere in the Age of the French Revolution* (Ithaca, N.Y., 1988); Dena Goodman, *The Republic of Letters: A Cultural History of the French Enlightenment* (Ithaca, N.Y., 1994); Hull, *Sexuality, State, and Civil Society*. For different interpretations that suggest a nuanced and complex picture in which the public sphere was more open to women, see, for example, Keith Baker, "Defining the Public Sphere in 18th-century France: Variations on a Theme by Habermas," in Calhoun, ed., *Habermas and the Public*, 201–7; Margaret C. Jacob, "The Mental Landscape of the Public Sphere: A European Perspective," *Eighteenth-Century Studies* 28, no. 1 (fall 1994): 95–113; Lawrence E. Klein, "Gender and the Public/Private Distinction in the Eighteenth Century: Some Questions about Evidence and Analytic Procedure," *Eighteenth-Century Studies* 29, no. 1 (fall 1995): 97–109. Finally, see also the insightful review of Isabel Hull's book by James Van Horn Melton in the *American Historical Review*, vol. 102, no. 5 (December 1997): 1509–10.

158. Marina Ledkovsky, Charlotte Rosenthal, and Mary Zirin, eds., *Dictionary of Russian Women Writers* (Westport, Conn., 1994), s.v. "Kheraskova, Elizaveta Vasil'evna"; "Kniazhnina, Ekaterina Aleksandrovna"; "Sushkova, Mar'ia Vasil'evna"; and "Urusova, Ekaterina Sergeevna." For more on Khrapovitskaia-Sushkova, who contributed to *All Sorts of Things*, the *Drone*, and *Evenings*, see Berkov, *Istoriia russkoi zhurnalistiki*, 228, 257, 292, 297.

159. The role of gender among leading eighteenth-century Russian male writers is discussed in Joe Andrew, *Women in Russian Literature, 1780–1863* (New York, 1988). On Svistunov, see D. D. Shamrai, "Ob izdateliakh pervogo chastnogo russkogo zhurnala (Po materialam arkhiva kadetskogo korpusa)," *XVIII vek* (Moscow-Leningrad, 1935), 379–81, 383–84. For another argument in favor of women's equality and their inclusion in society, see "Rassuzhdenie o priiatnostiakh soobshchestva," *Ezhemesiachnyia sochineniia k pol'ze i uveseleniiu sluzhashchie*, pt. 1 (February 1756): 153–79.

160. Information on these Masons comes from Serkov, "Rossiiskoe masonstvo." On the relations between the English Club and St. Petersburg lodges, see p. 58.

3. Virtue's Refuge, Wisdom's Temple

1. M. N. Longinov, *Novikov i moskovskie martinisty* (Moscow, 1867), 60.
2. OR RNB, f. 550, F.III.110, ll. 2ob, 5–5ob, 6ob.
3. Ibid., l. 7ob.
4. Ibid., F.III.111, l. 16.
5. OR RGB, f. 147, No. 123 /M.1999a/, ll. 10–10ob; OR RNB, f. 550, F.III.110, l. 4.
6. OR RNB, f. 550, F.III.111, l. 32.
7. Ibid., F.III.110, l. 7.
8. OPI GIM, f. 17, op. 2, d. 406, l. 2ob; N. S. Ivanina, ed. and trans., "K istorii masonstva v Rossii," *RS* 36 (October 1882): 72–73; OR RNB, f. 550, F.III.110, ll. 5, 11.
9. OR RNB, f. 550, F.III.110, ll. 13–13ob, 14.
10. Ibid., ll. 7ob–8.
11. Ibid., F.III.111, l. 24.
12. *Magazin svobodno-kamen'shchicheskoi*, vol. 1, pt. 1 (1784): 46, 48–49.
13. OR RNB, f. 550, F.III.110, l. 2. Although this specific reference comes from the bylaws of the Swedish-system Rising Sun lodge, I. P. Elagin's union of English-system lodges had the same provision. See G. V. Vernadskii, *Russkoe masonstvo v tsarstvovanie Ekateriny II* (Petrograd, 1917), 12.
14. OPI GIM, f. 398, d. 16, l. 20ob.
15. OR RNB, Titovskoe sobranie 4419, l. 26ob.
16. "Rech' vtoraianadesiat', o liubvi bratskoi. Govorennaia v nekotorom Khristianskom obshchestve v Anglii," *Moskovskoe ezhemesiachnoe izdanie*, pt. 3 (October 1781): 154.
17. From *L'Adoption ou la maçonnerie des femmes en trois grades* (n.p., 1775), 61. Quoted in Janet M. Burke and Margaret C. Jacob, "French Freemasonry, Women, and Feminist Scholarship," *Journal of Modern History* 68 (September 1996): 536.
18. OR RNB, f. 550, F.III.111, ll. 10ob, 11ob.
19. Ivanina, "K istorii masonstva," 36: 69; OR RNB, f. 550, F.III.111, ll. 15ob–16; F.III.110, l. 5.
20. OR RNB, f. 550, F.III.159, l. 2ob.
21. On Freemasonry as language (and, conversely, the language of Freemasonry), see Margaret C. Jacob, *Living the Enlightenment: Freemasonry and Politics in Eighteenth-Century Europe* (Oxford, 1991), 143–61; Marie Mulvey Roberts, "Masonics, Metaphor and Misogyny: A Discourse of Marginality?" in *Languages and Jargons: Contributions to a Social History of Language*, ed. Peter Burke and Roy Porter (Cambridge,

1995), 133–54. See also the interesting discussion of Russian Masonic literature that explores similar connections between language, Freemasons, and the unenlightened in Stephen L. Baehr, "Paradise Within: The Masonic Component of the Paradise Myth," chap. 5 in *The Paradise Myth in 18th-Century Russia: Utopian Patterns in Early Secular Russian Literature and Culture* (Stanford, 1991).

22. OR RGB, f. 14, No. 2, ll. 26–28.

23. OR RNB, f. 550, F.III.129, ll. 5ob, 8.

24. OR RGB, f. 14, No. 2, l. 28ob.

25. TsKhIDK, f. 1412, op. 1, d. 5300, l. 10.

26. ORRK BAN, no. 17.16.37, ll. 121–121ob. For a useful discussion of this and other Masonic songs, see A. V. Pozdneev, "Rannie masonskie pesni," *Scando-Slavica* 8 (1962): 26–64.

27. Mircea Eliade, *The Sacred and the Profane; the Nature of Religion* (New York, 1959), 184. The locus classicus on such transitional rituals is Arnold van Gennep, *The Rites of Passage* (1908; reprint, Chicago, 1960).

28. Ia. L. Barskov, *Perepiska moskovskikh masonov XVIII veka, 1780–1792 gg.* (Petrograd, 1915), 234; Vernadskii, *Russkoe masonstvo*, 69 n. 3.

29. A. I. Serkov, "Rossiiskoe masonstvo: Entsiklopedicheskii slovar', vol. 1, Masony v Rossii (1731–1799)," unpublished manuscript, Moscow, n.d, 666.

30. This description is based on the texts from the Rising Sun lodge (OR RNB, f. 550, F.III.111). For a published version of the standard initiation ceremony that corresponds with that conducted in Rising Sun, see *Iz bumag mitropolita moskovskago Platona* (Moscow, 1882), 1–27. See also Tira Sokolovskaia, "Obriadnost' vol'nykh kamenshchikov," in *Masonstvo v ego proshlom i nastoiashchem*, ed. S. P. Mel'gunov and N. P. Sidorov (1914–1915; reprint, Moscow, 1991), 2:86–95.

31. OR RNB, f. 550, Q.III.129, l. 5.

32. Ibid., F.III.111, l. 4ob.

33. Ibid., F.III.110, l. 7ob; ORRK BAN, no. 17.16.37, l. 121ob.

34. See Priscilla Roosevelt, *Life on the Russian Country Estate: A Social and Cultural History* (New Haven, 1995).

35. Capitals in original. RGADA, f. 8, op. 1, e./x. 255, l. 146. In the Apprentice initiation these three objects are listed as the "rough stone," the "cubical stone," and the "trestle-board." OR RNB, f. 550, F.III.111, ll. 16ob–17. Tira Sokolovskaia interprets these symbols in the following manner: "The rough stone is crude morality, chaos; the cubical stone is 'polished' morality; the trestle-board is a planned character of work, the authority of the good example." "Obriadnost'," in *Masonstvo v ego proshlom*, 2:98.

36. OR RNB, f. 550, F.III.113, l. 5ob.

37. Ibid., F.III.110, l. 14ob.

38. Ibid., ll. 2ob, 8ob–9, 10ob–11; Ivanina, "K istorii masonstva," 36: 70.

39. OR RNB, f. 550, F.III.112, l. 4ob.

40. Ibid., F.III.113, ll. 5ob, 18ob.

41. Ibid., F.III.111, l. 31ob; A. N. Pypin, *Russkoe masonstvo. XVIII i pervaia chetvert' XIX v.* (Petrograd, 1916), 47.

42. OPI GIM, f. 17, op. 2, d. 406, l. 80.

43. OR RNB, f. 550, F.III.110, l. 15.

44. OR RGB, f. 14, No. 2, ll. 51–52. This set of general laws contains a twelve-

point section on how to handle visitors.

45. OR RGB, f. 14, No. 2, ll. 51ob–52; OR RNB, f. 550, F.III.48, l. 1.

46. See Tira Sokolovskaia, "Masonskiia sistemy," in Mel'gunov and Sidorov, *Masonstvo v ego proshlom*, 2:52–79; Vernadskii, *Russkoe masonstvo*, 13–79; Pypin, *Russkoe masonstvo*, 103–35, 151–69, 218–62. Most students of Russian Freemasonry uncritically reiterated this very language of Masonry, that is, the concern with distinguishing "true" from "false" Freemasonry. Sokolovskaia's comment that various Russian Masonic systems sought to "return the immemorial purity and simplicity to Freemasonry . . . [by] preserving Masonic teaching in its greatest clarity" captures the inability of many past scholars to distance themselves from the internal debates that set the numerous Masonic groups against each other. "Masonskiia sistemy," in Mel'gunov and Sidorov, *Masonstvo v ego proshlom*, 2:78.

47. OPI GIM, f. 17, op. 2, d. 406, l. 4.

48. Vernadskii, *Russkoe masonstvo*, 19.

49. OPI GIM, f. 17, op. 2, d. 406, l. 61ob. See also Vernadskii, *Russkoe masonstvo*, 18 n. 8. Note, however, that the date given by Vernadskii (19 November 1774) does not correspond with that in the original document.

50. Reichel, a native German who had served at the court of the prince of Brunswick before coming to Russia, was a director of the Noble Corps of Cadets from 1771 to 1778.

51. Vernadskii, *Russkoe masonstvo*, 32–35; N. I. Novikov, *Izbrannye sochineniia*, ed. G. P. Makogonenko (Moscow-Leningrad, 1951), 607.

52. On the general history of the relationship between these two systems, see Vernadskii, *Russkoe masonstvo*, 19, 32–36; Pypin, *Russkoe masonstvo*, 119–35. For Urania's dealings with members of the Reichel system, see OPI GIM, f. 17, op. 2, d. 406, ll. 4ob, 5ob, 11, 21ob, 57–57ob, 80ob.

53. Ivanina, "K istorii masonstva," 36: 71.

54. P. Pekarskii, *Dopolneniia k istorii masonstva v Rossii XVIII stoletiia* (St. Petersburg, 1869), 118–19.

55. Within the system of Strict Observance, Andrew's Masonry referred to the fourth degree known as Scottish Master, usually rendered in Russian as *Shotlandskii gradus*. In the Swedish Rite and Rite of Zinnendorf, it referred to the fourth and fifth degrees. The fourth degree had three titles: Elect Master, Apprentice and Fellow Craft of St. Andrew, or Scotch Apprentice and Fellow Craft. The titles of the fifth degree were Scottish Master, Master of St. Andrew, or Grand Scottish Elect. For brief descriptions of these rites and their histories, see Robert Macoy, *A Dictionary of Freemasonry* (New York, 1989), 83–84, 318–20, 359–62, 366, 396.

56. OR RGB, f. 147, No. 123 /M. 1999a/, ll. 5–5ob.

57. Ibid., l. 2.

58. Ibid., ll. 5ob–6ob.

59. A similar practice in the Second Elagin Union governed the granting of the Royal Arch degree, which was limited to seventy-nine brothers.

60. OR RNB, f. 550, O.III.44, ll. 3–3ob.

61. Ibid., ll. 9–9ob, 13–15, 21ob, 27.

62. On the Theoretical Degree, see Vernadskii, *Russkoe masonstvo*, 65–73; Sokolovskaia, "Masonskiia sistemy," in Mel'gunov and Sidorov, *Masonstvo v ego proshlom*, 2:72–75. In Germany, the Theoretical Degree seems to have operated less as

an independent body and more as just another degree within the Rosicrucian order. On the Rosicrucians in Germany, see Christopher McIntosh, *The Rose Cross and the Age of Reason: Eighteenth-Century Rosicrucianism in Central Europe and Its Relationship to the Enlightenment* (Leiden, 1992). Also worthy of note is Gerhard Steiner's *Freimaurer und Rosenkreuzer: Georg Forsters Weg durch Geheimbünde* (Berlin, 1985).

63. OR RGB, f. 147, No. 100 /M. 1977/, l. 15.

64. Ibid., l. 3.

65. OR RNB, f. 550, F.III.47, ll. 77ob, 80–81.

66. OR RGB, f. 147, No. 100 /M. 1977/, l. 18.

67. Vernadskii, *Russkoe masonstvo*, 70.

68. OR RNB, f. 550, F.III.47, l. 45.

69. Ibid., F.III.48, ll. 19–19ob, 46ob, 118.

70. Ibid., F.III.47, l. 97.

71. On the Rosicrucians, see Vernadskii, *Russkoe masonstvo*, 65–79; Sokolovskaia, "Masonskiia sistemy," in Mel'gunov and Sidorov, *Masonstvo v ego proshlom*, 2:71–74.

72. Steiner, *Freimaurer und Rosenkreuzer*, 94; Vernadskii, *Russkoe masonstvo*, 72; OR RGB, f. 147, No. 294 /M 2168.2, l. 29.

73. OR RGB, f. 147, No. 294 /M. 2168.2/, ll. 29, 35, 57.

74. Ibid., ll. 38ob–39.

75. Ibid., /M. 2168.1/, ll. 1–4.

76. Written in French, Repnin's work was published under the title *Les Fruits de la grace* [sic], *ou Opuscules spirituels de deux amateurs de la Sagesse* (n.p., 1790). According to Vernadskii, another French edition was also published under a different title. *Russkoe masonstvo*, 261–62. The sections quoted above are from the book's first appendix entitled "Quelques avis historiques sur les F. M.," pp. 98–100. It is not entirely clear whether Repnin composed these two addenda as well as the main text.

77. Novikov, *Izbrannye*, 609–11, 627. Vernadskii lists some sixty Rosicrucians active in Russia. *Russkoe masonstvo*, 272–80.

78. Novikov's words, especially those on the nature of English Masonry, which is usually seen as representative of all early Masonry in Russia, are repeated, and taken as an accurate characterization, in practically every work on the subject. See, for example, Vernadskii, *Russkoe masonstvo*, 20–21.

79. Wilson R. Augustine, "Notes Toward a Portrait of the Eighteenth-Century Russian Nobility," *Canadian Slavic Studies* 4, no. 3 (fall 1970): 375–76; Marc Raeff, *Origins of the Russian Intelligentsia: The Eighteenth-Century Russian Nobility* (New York, 1966), 111–12, 118; idem, *Understanding Imperial Russia: State and Society in the Old Regime*, trans. Arthur Goldhammer (New York, 1984), 41; Elise Kimerling Wirtschafter, *Structures of Society: Imperial Russia's "People of Various Ranks"* (DeKalb, Ill., 1994), 74–76.

80. *Izbrannye proizvedeniia*, ed. L. I. Kulakova (Leningrad, 1961), 318.

81. See S. O. Shmidt, "Obshchestvennoe samosoznanie Noblesse russe v XVI–pervoi treti XIX v.," *Cahiers du monde russe et soviétique* 34, nos. 1–2 (January–June 1993): 11–32. And for a contemporary statement, see Kniazhnin's "Poslanie k rossiiskim pitomtsam svobodnykh khudozhestv," in *Izbrannoe*, ed. A. P. Valagin (Moscow, 1991), 332–34.

82. "Zapiski A. T. Bolotova," *RS* 10, supplement (1872): 933.

83. Raeff, *Origins*, 161.

84. This interpretation of Masonic secrecy shapes not only the literature on Russian Masonry, but also much of the literature on the movement in other parts of Europe. See, for example, the influential work of Reinhart Koselleck, *Critique and Crisis: Enlightenment and the Pathogenesis of Modern Society* (1959; English trans., Cambridge, Mass., 1988). The origins of the view of secrecy as a shield against the state can be traced back to Kant. See "On the Common Saying: 'This May be True in Theory, but it does not Apply in Practice,'" in *Kant's Political Writings*, ed. Hans Reiss, trans. H. B. Nisbet (Cambridge, 1970), 85–86.

85. V. Bogoliubov, *N. I. Novikov i ego vremia* (Moscow, 1916), 180; Vernadskii, *Russkoe masonstvo*, 5–9. Bogoliubov points out the inaccuracies in I. V. Böber's oft-quoted statements concerning the supposed dangers facing Russian Masons under Elizabeth and the need to adopt extreme measures to ensure their safety. For examples of the information on Freemasonry supplied to the political police in the reign of Elizabeth, see *Letopisi russkoi literatury i drevnosti* 4, sect. 3 (1862): 49–52. A recent study of Masonry by a Russian historian makes a similar observation about the tolerant atmosphere under Elizabeth I. See O. F. Solov'ev, *Russkoe masonstvo, 1730–1917* (Moscow, 1993), 33.

86. Ernst Friedrichs, *Geschichte der einstigen Maurerei in Rußland* (Berlin, 1904), 18; Longinov, *Novikov*, 101, 110 n. 61; Pekarskii, *Dopolneniia*, 3–4; Vernadskii, *Russkoe masonstvo*, 9–10, 56. The participant was the young Mason A. Ia. Il'in. OR RNB, f. 487, op. 2, ch. 2, e./x. O.87, l. 231ob.

87. Lopukhin, *Zapiski senatora I. V. Lopukhina* (1859; reprint, Moscow, 1990), 57.

88. Longinov, *Novikov*, 110; Vernadskii, *Russkoe masonstvo*, 47, 227. According to Vernadskii, Lopukhin visited Gagarin's National lodge on two separate occasions. Throughout her reign, Catherine II was especially distrustful of Swedish Freemasonry and its alliance with Russia's lodges. Following Lopukhin's inspection of the lodges allied with Sweden, Catherine appears to have ordered them shut down and to have repeated her directive against all Swedish-system lodges in the late 1780s, after their successful revival several years earlier. N. S. Ivanina, "K istorii masonstva v Rossii," *RS* 35 (September 1882): 542–43.

89. Georg Reinbeck, *Travels from St. Petersburgh through Moscow, Grodno, Warsaw, Breslaw, etc. to Germany, in the Year 1805*, in *A Collection of Modern and Contemporary Voyages and Travels*, ed. Richard Phillips, vol. 6 (London, 1807), 127–30.

90. Friedrichs, *Geschichte*, 64–65. Unfortunately Friedrichs does not say who these friends of Böber were. It should be noted that this is the same I. V. Böber whose characterization of Freemasonry under Elizabeth was referred to above as less than reliable (see n. 85). Thus, the story about Grenet must be regarded with a healthy degree of skepticism.

91. Longinov, *Novikov*, 246; Novikov, *Izbrannye*, 658–59.

92. Ivanina, "K istorii masonstva," 35: 542; Vernadskii, *Russkoe masonstvo*, 36–37.

93. Vernadskii, *Russkoe masonstvo*, 64.

94. See also the noted memoirist A. T. Bolotov's account of Novikov's repeated attempts to entice him to join the movement. "Zapiski A. T. Bolotova," *RS* 10, supplement (1872): 931–35; *RS* 12, supplement (1872): 1134. Serkov lists Bolotov as having joined the Freemasons while abroad. "Rossiiskoe masonstvo," 239–41. Also

worth mentioning are the reminiscences of Baron de Corberon, France's chargé d'affaires in St. Petersburg and an active figure in both capitals' Masonic circles. Marie Daniel Bourrée, baron de Corberon, *Un diplomate français à la cour de Catherine II, 1775–1780: Journal intime,* ed. L. H. Lablande (Paris, 1901), 1:106–8, 161; 2:3–4, 139, 175, 395–96.

95. All three of them belonged to the Equality lodge, founded in September 1774, that met during 1775 in the home of Prince M. M. Shcherbatov in Krasnoe Selo outside Moscow. OR RNB, f. 487, op. 2, ch. 2, e./x. O.87, l. 72; Serkov, "Rossiiskoe masonstvo," 862–66.

96. OR RNB, f. 487, op. 2, ch. 2, e./x. O.87, ll. 22, 70, 72–72ob.

97. Ibid., l. 72ob.

98. Ibid., ll. 164, 172ob–173, 227–227ob, 247. See also V. I. Savva, "Iz dnevnika masona 1775–1776 gg.," *Chteniia v Imperatorskom Obshchestve istorii i drevnostei rossiiskikh* 4 (1908): 10–15. Note that Savva's page citations from Il'in's diary do not correspond with my own as the manuscript contains two different paginations. I chose the one that seemed more accurate.

99. OR RNB, f. 487, op. 2, ch. 2, e./x. O.87, ll. 172, 209–209ob.

100. OR RNB, f. 550, F.III.110, l. 8ob.

101. The connection between veiling and marking spaces as sacred is addressed in van Gennep, *Rites,* 168.

102. *The Presentation of the Self in Everyday Life* (New York, 1959), 142. Georg Simmel made a similar observation much earlier: "The separateness of the secret society expresses a value: people separate from others because they do not want to make common cause with them, because they wish to let them feel their superiority." *The Sociology of Georg Simmel,* ed. and trans. Kurt H. Wolff (Glencoe, Ill., 1950), 364. See as well the excellent article by Wolfgang Hardtwig, "Eliteanspruch und Geheimnis in den Geheimgesellschaften des 18. Jahrhunderts," in *Aufklärung und Geheimgesellschaften: Zur politischen Funktion und Sozialstruktur der Freimaurerlogen im 18. Jahrhundert,* ed. Helmut Reinalter (Munich, 1989), 63–86.

103. Simmel notes that there is a strong attraction to secrecy "beyond its significance as a mere means"—that is, irrespective of the "contents" it is hiding. Rather, the attraction comes from the fact that "the strongly emphasized exclusion of all outsiders makes for a correspondingly strong feeling of possession." *Sociology of Georg Simmel,* 332.

104. Lopukhin, *Zapiski senatora I. V. Lopukhina,* 23.

105. Simmel, *Sociology of Georg Simmel,* 349–50.

106. OR RNB, f. 550, O.III.32, l. 1.

107. Ibid., F.III.111, ll. 15ob-16.

108. Mullan, *Sentiment and Sociability: The Language of Feeling in the Eighteenth Century* (New York, 1988), 124.

109. Tuchkov, *Zapiski* in *Zolotoi vek Ekateriny Velikoi: Vospominaniia,* ed. V. M. Bokova and N. I. Tsimbaev (Moscow, 1996), 173–74.

110. Pypin, *Russkoe masonstvo,* 275.

111. OR RNB, f. 550, F.III.112, l. 4.

112. Ibid., F.III.113, ll. 4–4ob.

113. OR RNB, f. 487, op. 2, ch. 2, e./x. O.87, ll. 85–85ob; Savva, "Iz dnevnika

masona," 2. Tred'iakovskii was the son of V. K. Tred'iakovskii, the well-known poet, playwright, and literary theoretician.

114. OR RNB, f. 550, F.III.110, l. 10ob.

115. Bourrée, *Un diplomate français*, 1:107.

116. [Wegelin], "Pis'mo neizvestnago litsa o moskovskom masonstve XVIII veka," *RA*, bk. 1 (1874): 1033–34.

117. OR RNB, f. 550, F.III.111, l. 25; F.III.112, l. 12.

118. Ibid., F.III.113, ll. 10ob-13. For a description of the Master Mason initiation ceremony and the story of Hiram Abiff, occasionally also called Adoniram, see Sokolovskaia, "Obriadnost'," in Mel'gunov and Sidorov, *Masonstvo v ego proshlom*, 2:99–103. See also James Stevens Curl, *The Art and Architecture of Freemasonry* (London, 1991), 32–34.

119. OR RNB, f. 550, F.III.113, l. 18ob. Upon receiving the apron as a new Fellow Craft Mason, the brother is told only that "You will learn the meaning of the three white roses on your apron in the future when your masters think it fit to explain it to you." In other words, he must wait until having been deemed worthy of becoming a Master Mason. Ibid., F.III.112, l. 4ob.

120. Ibid., F.III.111, l. 25; F.III.112, l. 12ob; F.III.113, l. 19ob. The reference to the three thresholds most likely refers to the three gates before the temple, which led from the Court of the Gentiles, to the Court of the Children of Israel, to the Court of the Priests just before the twelve steps ascending to the temple's porch. Most Masonic discussions of the temple divide the structure into three parts: porch, holy place (or sanctuary), and holy of holies. Although the texts exhibit a great deal of variation in their depictions of the temple and not all assign the same location to the three degrees, they all imply a direct correlation between the highest degree and the temple's innermost spaces. For an informative discussion of the temple's history in Masonic lore, see Curl, *Art and Architecture*, 80–104.

121. Albert G. Mackey, *An Encyclopedia of Freemasonry* (Philadelphia, 1874), 154; Macoy, *Dictionary*, 105–6.

122. OR RNB, f. 550, F.III.113, l. 17.

123. Vernadskii, *Russkoe masonstvo*, 29.

124. OR RGB, f. 147, No. 90 /M. 1967/, l. 15.

125. OR RNB, f. 550, O.III.44, l. 28ob.

126. OR RGB, f. 147, No. 123 /M. 1999v/, l. 3. This document is part of a collection of three red notebooks (numbered 1999a, 1999b, 1999v) containing the official texts from a Scottish lodge active in the late 1780s under the Grand Master Iosef Pozdeev.

127. Ibid., /M. 1999a/, l. 8.

128. [Joseph Friedrich Göhring], *Dolzhnosti brat'ev Z. R. K. drevniia sistemy, govorennyia Khrizoferonom, v sobraniiakh iunioratskikh, s prisovokupleniem nekotorykh drugikh rechei drugikh brat'ev* (Moscow, 1784), 94–95. Originally published in German, this work was composed by the Rosicrucian Göhring in 1782 in response to *Über Jesuiten, Freymaurer und deutsche Rosencreutzer*, Friedrich Adolph Freiherr von Knigge's anti-Rosicrucian tract of 1781. Steiner, *Freimaurer und Rosenkreuzer*, 96–97.

129. Göhring quoted in Steiner, *Freimaurer und Rosenkreuzer*, 96–97.

130. Vernadskii, *Russkoe masonstvo*, 56–57

131. Ibid., 41–44. A Masonic official in Sweden wrote Kurakin that the rank-and-file members "do not know . . . and must not know" anything about this

supreme body. G. P. Gagarin, the chapter's head, pledged that only the "most reliable followers of the newly introduced system, the elite, enlightened brothers" would ever learn of this "Invisible Chapter."

132. Ivanina, "K istorii masonstva," 35: 539–41; Vernadskii, *Russkoe masonstvo,* 58–64.

133. Barskov, *Perepiska,* 260; Vernadskii, *Russkoe masonstvo,* 51–53; Longinov, *Novikov,* 147. In-Ho L. Ryu points out that not only did the Theoretical Brothers receive secret names, but they abandoned them for new ones upon attaining full membership among the Rosicrucians. Novikov, for example, changed his code name from "eq. ab. ancora" to "Colovion" after joining the Rosicrucians. "Moscow Freemasons and the Rosicrucian Order: A Study in Organization and Control," in *The Eighteenth Century in Russia,* ed. J. G. Garrard (Oxford, 1973), 220.

134. David Stevenson, *The Origins of Freemasonry: Scotland's Century, 1590–1710* (Cambridge, 1988), 96–105; Frances A. Yates, *The Rosicrucian Enlightenment* (London, 1972).

135. Bourrée, *Un diplomate français,* 2:3–4, 139, 175.

136. OR RGB, f. 147, No. 100 /M. 1977/, ll. 3–4.

137. OR RNB, f. 550, F.III.47, ll. 158–59. See also Wegelin's dissatisfaction with his "invisible superiors" in "Pis'mo neizvestnago litsa," 1037–38.

138. TsKhIDK, f. 1412, op. 1, d. 5300, l. 12. This file contains numerous letters sent from Stockholm to Kurakin between 1777 and 1779 that are particularly revealing for the light they shed on the introduction of Swedish Freemasonry into Russia. Among other things, Kurakin's letter suggests that Elagin refused the position of Grand National Master not out of fear of incurring Catherine's displeasure, as Vernadskii posited, but because he accurately sensed Kurakin's motives. Vernadskii, *Russkoe masonstvo,* 40.

139. Charles François Philibert Masson, *Secret Memoirs of the Court of Petersburg* (1802; reprint, New York, 1970), 99–100. See also Iu. M. Lotman, *Besedy o russkoi kul'ture: Byt i traditsii russkogo dvorianstva (XVIII–nachalo XIX veka)* (St. Petersburg, 1994), 101.

140. Simmel, *Sociology of Georg Simmel,* 360.

141. On the all-seeing eye, see Mackey, *Encyclopedia,* 57.

142. OR RNB, f. 550, F.III.47, l. 159.

143. Longinov, *Novikov,* 160, 199.

144. Ryu, "Moscow Freemasons," 218–21; Vernadskii, *Russkoe masonstvo,* 76.

145. OR RGB, f. 147, No. 294 /M. 2168.2/, ll. 27–28. The word "gatherings" has been used to translate the symbol "OOgi" in the original text. "Gatherings" appears to convey the sense, if not necessarily the exact meaning, of this sign.

146. Simmel, *Sociology of Georg Simmel,* 372.

147. Vernadskii, *Russkoe masonstvo,* 76–77.

148. For an excellent example, see the excerpts from the Masonic song "Ostavi svoi chertog" in Pozdneev, "Rannie masonskie pesni," 55.

149. OR RNB, f. 550, F.III.111, ll. 12ob–13ob.

150. On the "internal man," see Vernadskii, *Russkoe masonstvo,* 137–48. See also the speeches from the lodge in Orel: OR RNB, f. 550, F.III.47, ll. 36ob–47; F.III.48, ll. 103–6.

151. OR RNB, f. 550, O.III.123, O.III.124, O.III.159, O.III.160. Several of Gamaleia's speeches from the Deucalion lodge were also published in *Magazin svobodno-kamen'shchicheskoi.* See Vernadskii, *Russkoe masonstvo,* 257–58.

152. OR RGB, f. 147, No. 294 /M. 2168.2/, ll. 40, 57ob.

153. OR RNB, f. 550, F.III.111, l. 11ob.

154. Ibid., F.III.110, l. 5ob.

155. OR RGB, f. 147, No. 100 /M. 1977/, l. 3; No. 294 /M. 2168.2/, l. 35.

156. S. V. Eshevskii, *Sochineniia po russkoi istorii* (Moscow, 1900), 214. Ferdinand's position in European Masonry is addressed in Ludwig Hammermayer, *Der Wilhelmsbader Freimaurer-Konvent von 1782*, Wolfenbütteler Studien zur Aufklärung, Band 5/2 (Heidelberg, 1980).

157. Eshevskii, *Sochineniia*, 214.

158. On these events, see Vernadskii, *Russkoe masonstvo*, 51–54.

159. Barskov, *Perepiska*, 238–39. Trubetskoi's letters illustrate yet another Masonic distinction between inside and outside, according to which the former refers to the order's content (that is, teachings and mysteries) and the latter to its form (that is, organizational structure).

160. OR RGB, f. 147, No. 295 /M. 2169/, l. 41.

161. In fact, the three degrees of St. John's Masonry are generally referred to as the "symbolic degrees." Mackey, *Encyclopedia*, 782–83.

162. OR RNB, f. 550, Q.III.129, ll. 2–4ob.

163. Ibid., Q.III.153, l. 52ob. Only a portion of the entire chart is reproduced here. The exact format and order of the entries as contained in the original, including the lines separating the cells, have been kept in this reproduction. The vertical lines dividing the three columns mark the primary boundaries of secrecy within the Masonic order.

164. OR RNB, f. 550, F.III.111, ll. 11–11ob.

165. Although *tainstvo* is usually translated today as "sacrament," earlier meanings included "mystery" or "secret."

166. OR RNB, f. 550, F.III.111, l. 10ob.

167. [Wegelin], "Pis'mo neizvestnago litsa," 1033.

168. On Melissino, see Ryu, "Moscow Freemasons," 202 n. 8. On the Moscow Masons, see Vernadskii, *Russkoe masonstvo*, 126–30. A good deal of their mysteries were contained in the official texts of the Theoretical Degree. See OR RGB, f. 147, No. 100 /M. 1977/, ll. 38–110.

169. Vernadskii, *Russkoe masonstvo*, 133–37. For a more thorough description of Elagin's collection of mystical writings, see Pekarskii, *Dopolneniia*, 51–59, 92–115. Just as the mysteries of Melissino's system and those of the Moscow Masons were intended only for the most elite brothers, the writings in Elagin's "Teachings of Ancient Secular and Spiritual Wisdom" were meant to be read only to members of the secret governing body of his Second Elagin Union.

170. Bourrée, *Un diplomate français*, lxiv–lxv; Robert Darnton, *Mesmerism and the End of the Enlightenment in France* (Cambridge, Mass., 1968). On Corberon's links to Mesmerism, see pp. 76–77, 116, 180–82. Isaiah Berlin, *The Magus of the North: J. G. Hamann and the Origins of Modern Irrationalism* (New York, 1993).

171. E. Tarasov, "K istorii russkago obshchestva vtoroi poloviny XVIII stoletiia: Mason I. P. Turgenev," *Zhurnal Ministerstva narodnago prosveshcheniia*, n.s., 51, no. 6 (1914): 156–57.

172. "The Forms of Wildness: Archaeology of an Idea," in *The Wild Man Within: An Image in Western Thought from the Renaissance to Romanticism*, ed. Edward

Dudley and Maximillian E. Novak (Pittsburgh, 1972), 4–5.

173. For a useful discussion of the idea of societies' "active centers," see Liah Greenfeld and Michel Martin, eds., *Center: Ideas and Institutions* (Chicago, 1988).

174. On the contradictory nature of societal margins, see Mary Douglas, *Purity and Danger: An Analysis of the Concepts of Pollution and Taboo* (London, 1966), esp. 94–113, 121; Victor Turner, *The Ritual Process: Structure and Anti-Structure* (Ithaca, N.Y., 1969), esp. 108–11.

4. THE IMAGE OF THE MASON

1. A. N. Pypin, *Russkoe masonstvo. XVIII i pervaia chetvert' XIX v.* (Petrograd, 1916), 97.

2. M. N. Longinov, *Novikov i moskovskie martinisty* (Moscow, 1867), 93.

3. G. R. Derzhavin, *Sochineniia*, ed. Ia. Grot, vol. 6 (St. Petersburg, 1871), 437–38.

4. N. Popov, "Pridvornyia propovedi v tsarstvovanie Elisavety Petrovny," *Letopisi russkoi literatury i drevnosti* 2, bk. 3 (1859): 1–32. The quotes are on p. 22. V. Bogoliubov, *N. I. Novikov i ego vremia* (Moscow, 1916), 179; Pypin, *Russkoe masonstvo,* 98; T. Sokolovskaia, *Russkoe masonstvo i ego znachenie v istorii obshchestvennogo dvizheniia (XVIII i pervaia chetvert' XIX stoletiia)* (St. Petersburg, [1907]), 7.

5. Pypin, *Russkoe masonstvo,* 91; "Donesenie o masonakh," *Letopisi russkoi literatury i drevnosti* 4, sect. 3 (1862): 51–52. See also in the same publication (pp. 49–51) Shuvalov's interrogation of the Mason Mikhail Olsuf'ev from the mid-1750s.

6. Pypin, *Russkoe masonstvo,* 98–99.

7. I. F. Martynov, "Rannie masonskie stikhi i pesni v sobranii Biblioteki Akademii nauk SSSR (k istorii literaturno-obshchestvennoi polemiki 1760-kh gg.)," in *Russia and the World of the Eighteenth Century,* ed. R. P. Bartlett, A. G. Cross, and Karen Rasmussen (Colombus, Ohio, 1988), 437–39. See also idem, "Masonskie rukopisi v sobranii Biblioteki AN SSSR" in *Materialy i soobshcheniia po fondam otdela rukopisnoi i redkoi knigi BAN SSSR* 2 (Leningrad, 1978), 243–53.

8. Martynov, "Rannie masonskie stikhi," 439; Pypin, *Russkoe masonstvo,* 96–98. The copy of the text used here is in TsNB-KRK GE, Inv. No. 150554, ll. 230ob–233. The Russian title reads "Iz"iasnenie nekotorykh izvestnykh del prokliatago zborishcha frank mazonskago."

9. TsNB-KRK GE, Inv. No. 150554, l. 230ob.

10. Ibid., ll. 230ob–23lob.

11. Numerous parodies of Freemasonry as a world inverted in which vice has supplanted virtue, darkness light, and so on, exist from this period. For an interesting discussion, see A. V. Pozdneev, "Rannie masonskie pesni," *Scando-Slavica* 8 (1962): 26–64.

12. TsNB-KRK GE, Inv. No. 150554, ll. 231–232.

13. Ibid., l. 231–231ob.

14. Ibid., l. 232ob.

15. Ibid., l. 233.

16. Ibid., ll. 231–233.

17. It has been suggested that "A Psalm Exposing the Freemasons" ("Psal'ma, na oblichenie frank-masonov") was composed as a direct response to I. P. Elagin's

satire "On a Dandy and a Coquette" (1753), which launched a round of social-literary polemics among the country's leading *gens de lettres*. Elagin's opponents in this literary war were likely aware of his—and his ally A. P. Sumarokov's—strong Masonic affiliation and used this as a way to score points in these debates. For a discussion of these polemics, see I. F. Martynov and I. A. Shanskaia, "Otzvuki literaturno-obshchestvennoi polemiki 1750-kh godov v russkoi rukopisnoi knige. (Sbornik A. A. Rzhevskogo)," *XVIII vek*, vol. 11 (Leningrad, 1976), 131–48. On the origins of the psalm, see Martynov, "Rannie masonskie stikhi," 439.

18. *Bibliograficheskiia zapiski* 3, no. 3 (1861): 70. The history of the particular manuscript from which this copy of the psalm was published is discussed in Martynov, "Masonskie rukopisi," 249 n. 9.

19. The "Declaration" mentions Freemasons singing "Parisian songs" and contains almost the exact couplet found in the "Psalm": "The French for 'maçon' / Is none other than 'mason' in Russian." TsNB-KRK GE, Inv. No. 150554, l. 231.

20. For an intriguing article that claims to uncover in the Gallophobia of the day the origins of the idea of an international Masonic plot directed against Russia, see Andrei Zorin, "K predystorii odnoi global'noi kontseptsii (Oda V. P. Petrova "Na zakliuchenie s Ottomanskoiu Portoiu mira" i evropeiskaia politika 1770-kh godov)," *Novoe literaturnoe obozrenie* 23 (1997): 56–77.

21. N. G. Kurganov, *Rossiiskaia universal'naia grammatika, ili Vseobshchee pismoslovie, predlagaiushchee legchaishii sposob osnovatel'nago ucheniia ruskomu* [sic] *iazyku s sedm'iu prisovokupleniiami raznykh uchebnykh i poleznozabavnykh veshchei* (St. Petersburg, 1769). The version published in Kurganov's text differs in a few minor ways from the manuscript version above.

22. On Kurganov and his book, see I. E. Barenbaum, *Istoriia knigi,* 2nd ed. (Moscow, 1984), 55; T. V. Staniukovich, *Kunstkamera peterburgskoi Akademii nauk* (Moscow-Leningrad, 1953), 164; A. V. Zapadov, "Iv. Novikov, Komarov, Kurganov," in *Istoriia russkoi literatury,* vol. 4 (Moscow-Leningrad, 1947), 308–10.

23. According to A. V. Pozdneev, one of the major themes of early Masonic songs was the defending of the order against its numerous detractors. "Rannie masonskie pesni," 39–43.

24. ORRK BAN, No. 17.16.37, ll. 120–20ob.

25. On the origins of Sumarokov's defense, see Martynov, "Rannie masonskie stikhi," 439–40. The work was first published in Sumarokov's *Polnoe sobranie vsekh sochinenii,* pt. 8 (Moscow, 1781), 323–34, although it had circulated in manuscript long before its publication. See Pozdneev, "Rannie masonskie pesni," 48. For a slightly different translation, see William Edward Brown, *A History of 18th-Century Russian Literature* (Ann Arbor, Mich., 1980), 162. On the poet's reputation in Russia, see Simon Karlinsky, *Russian Drama from Its Beginnings to the Age of Pushkin* (Berkeley-Los Angeles, 1985), 70–71, 100–101.

26. N. K. Piksanov, "Masonskaia literatura," in *Istoriia russkoi literatury,* vol. 4 (Moscow-Leningrad, 1947), 63.

27. Steven C. Bullock, *Revolutionary Brotherhood: Freemasonry and the Transformation of the American Social Order, 1730–1840* (Chapel Hill, N.C., 1996), 80–81; Margaret C. Jacob, *Living the Enlightenment: Freemasonry and Politics in Eighteenth-Century Europe* (New York, 1991), 5–6, 23–28, 76–77, 152–53, 168; J. M. Roberts, *The Mythology of the Secret Societies* (London, 1972), 68–84, 85.

28. *Obol'shchennyi, komediia v piati deistviiakh,* in *Sochineniia Imperatritsy Ekateriny II,* ed. A. N. Pypin, vol. 1 (St. Petersburg, 1901), Act I, Scene 1.

29. Act I, Scene 3.

30. Act II, Scene 1.

31. Act I, Scene 1.

32. Act IV, Scenes 1–6; Act V, Scene 1.

33. Act V, Scene 8.

34. Act V, Scenes 2, 13.

35. Act V, Scene 15. Ellipses in original.

36. On Catherine's anti-Masonic plays, see also A. V. Semeka, "Russkie rozenkreitsery i sochineniia imperatritsy Ekateriny II protiv masonstva," *Zhurnal Ministerstva narodnago prosveshcheniia,* pt. 39, no. 2 (1902): 343–400.

37. *The Deceiver* was staged most often—over a dozen times. While this number may at first appear small, most plays in eighteenth-century Russia received an average of approximately ten stagings. By comparison, one of the most often performed works of the century, Aleksandr Ablesimov's comic opera *The Miller: A Wizard, a Cheat, and a Matchmaker (Mel'nik—koldun, obmanshchik i svat)* (1779), was staged over eighty times in Moscow and St. Petersburg. See V. N. Vsevoldskii-Gerngross, *Istoriia russkogo dramaticheskogo teatra,* vol. 1, *Ot istokov do kontsa XVIII veka* (Moscow, 1977), 451–52, 456. *The Deceiver* was also performed in German in St. Petersburg and Hamburg. See the commentary to these plays by A. N. Pypin in *Sochineniia Imperatritsy Ekateriny II,* vol. 1. According to one source, Catherine's plays met with great success among the St. Petersburg theater-going public. Longinov, *Novikov,* 256–58.

38. Longinov, *Novikov,* 258.

39. Bogoliubov, *Novikov i ego vremia,* 364. For information on the comedies' Russian editions, see *SK,* vol. 1 (Moscow, 1962), nos. 2141–44, 2146. The German journalist C. F. Nicolai published German translations of the plays in Berlin in 1788. (Pypin, *Russkoe masonstvo,* 310). For information on the writing, staging, and publication of Catherine's plays, see also the commentary included in *Sochineniia Imperatritsy Ekateriny II,* vol. 1.

40. *Zerkalo sveta* pt. 1, no. 1 (9 February 1786): 13–15.

41. *Rastushchii vinograd* (February 1786): 1–6. Italics in original.

42. Ibid., 43–44. Later that year, the journal ran a lengthy article (clearly a translation) on the origins of Freemasonry: "Razsuzhdenie o nachale vol'nykh kamenshchikov ili Farmasonov" (October 1786): 44–72; (December 1786): 62–76; (January 1787): 35–46. Echoes of the reactions to Catherine's plays can also be found in *Novyia ezhemesiachnyia sochineniia,* pt. 2 (August 1786): 70–82; pt. 3 (September 1786): 76–77.

43. Vsevolodskii-Gerngross, *Istoriia russkogo dramaticheskogo teatra,* 1:452. Emin's play was published anonymously by the Academy of Sciences the following year. *Mnimyi mudrets: Komediia v piati deistviiakh* (St. Petersburg, 1786). Emin was the author of several comedies written and performed in the 1780s and 1790s.

44. Emin, *Mnimyi mudrets,* 8, 27.

45. Ibid., 22.

46. Ibid., 3, 8–10, 52.

47. This point is appropriately made in Karlinsky, *Russian Drama,* 86–87, 91.

48. In a letter of January 1786 to the Swiss doctor and writer J. G. Zimmer-

mann, Catherine writes that *The Deceiver* depicts Cagliostro and *The Deceived* portrays those whom he has deluded. *Filosoficheskaia i politicheskaia perepiska Imperatritsy Ekateriny II s Doktorom Tsimmermanom, s 1785 po 1792 god* (St. Petersburg, 1803), 45–46. See also Zimmermann's response on pp. 50–53. The Russian public appears not to have had any difficulty in figuring out Catherine's target. In a letter also of January 1786 to a friend abroad, Prince P. D. Tsitsianov writes about how *The Deceiver* mocks "those who believe in charlatans and especially believed in Cagliostro." "Pis'ma kniazia Pavla Dmitrievicha Tsitsianova k Vasiliiu Nikolaevichu Zinov'evu," *RA*, bk. 2 (1872): 2113.

49. For an overview of Cagliostro's spectacular life and adventures, see Roberto Gervaso, *Cagliostro: A Biography*, trans. Cormac Ó Cuilleanáin (London, 1974). On Cagliostro's stay in Courland and Russia, compare the accounts in E. P. Karnovich, *Zamechatel'nyia i zagadochnyia lichnosti XVIII i XIX stoletii*, 2nd ed. (St. Petersburg, 1893), 98–124; Baron Karl-Heinrich Heyking, "Vospominaniia senatora barona Karla Geikinga," *RS* 91 (September 1897): 532–35; Peter Wilding, *Adventurers in the Eighteenth Century* (New York, [1937]), 296–302; V. Zotov, "Kaliostro, ego zhizn' i prebyvanie v Rossii," *RS* 12 (January 1875): 50–83. Cagliostro's own treatment of his stay in Russia can be found in *Confessions du comte de C*** avec l'histoire de ses voyages en Russie, Turquie, Italie, et dans les pyramides d' Egpyte* (Au Caire [Paris], 1787), 55–66.

50. Despite Karnovich's statements to the contrary, Cagliostro appears to have lived with Elagin during his stay in Russia. There is a notice announcing the departure of Cagliostro and his wife from the capital in the *St. Petersburg News*, 21 February 1780, no. 15, *Pribavlenie*, 191 (repeated in no. 16, 204) that states the couple is living in the home of I. P. Elagin. See also Catherine's humorous description of Cagliostro's stay in Elagin's residence in a letter to Baron F. M. Grimm. "Pis'ma Imperatritsy Ekateriny II k Grimmu (1774–1796)," *Sbornik Imperatorskago russkago istoricheskago obshchestva* 23 (1878): 212–13.

51. Elagin's manuscripts contain eighteen prescriptions that he received from Cagliostro. RGADA, f. 8, op. 1, e./x. 238, ll. 51–62. This file also contains various numerological graphs and charts as well as directions for guessing correct lottery numbers, a knowledge that Cagliostro professed and that he shared, for a price, with others. See ll. 1–24, and François Ribadeau Dumas, *Cagliostro*, trans. Elisabeth Abbott (New York, 1967), 69–72.

52. Compare the two different versions of this episode in Ernst Friedrichs, *Geschichte der einstigen Maurerei in Rußland* (Berlin, 1904), 80, and Wilding, *Adventurers*, 300.

53. Heyking, "Vospominaniia," 533–34; Zotov, "Kaliostro," 64–65.

54. Karnovich, *Zamechatel'nyia i zagadochnyia lichnosti*, 114–15; Wilding, *Adventurers*, 300–301.

55. Karnovich, *Zamechatel'nyia i zagadochnyia lichnosti*, 119.

56. On the affair, see Wilding, *Adventurers*, 312–18. The count's works were published under the titles *Memorial grafa Kalliostro protiv gospodina general prokurora obviniaiushchago ego, pisannoi im samim* (Moscow, 1786); *Opravdanie grafa de Kalliostro po delu kardinala Rogana o pokupke slavnago sklavazha vo Frantsii* (St. Petersburg, 1786). See *SK*, vol. 2 (Moscow, 1964), nos. 2754 and 2755. The fact that his *Treatise* was translated and published in the very year it appeared in France suggests the impor-

tance and sense of urgency Cagliostro's Russian defenders felt about their cause.

57. [Doillot], *Vozrazhenie so storony grafiny de Valua–la Mott, na opravdanie grafa de Kalliostro* (St. Petersburg, 1786); *Opisanie prebyvaniia v Mitave izvestnago Kaliostra na 1779 god, i proizvedennykh im tamo magicheskikh deistvii* (St. Petersburg, 1787), 35–36, 63. This work was first published in German that same year in Berlin. The author, who had herself originally been fooled by Cagliostro and whose family had hosted the Grand Copt in Mitau, attempted to establish a hidden Jesuit connection behind the actions of Cagliostro and his ilk, seeing in them the machinations of the order. For more on Charlotte von der Recke and her text, see Pypin, *Russkoe masonstvo*, 284–87. A. F. Moshchinskii [Moszynski], *Kalliostr poznannyi v Varshave, ili Dostovernoe opisanie khimicheskikh i magicheskikh ego deistvii, proizvodimikh v sem stolichnom gorode v 1780* (Moscow, 1788). A friend of Prince Poninski, in whose Warsaw home Cagliostro stayed after leaving Russia, Moszynski initially sought to work as the Italian's laboratory assistant, before becoming convinced of his duplicity. Wilding, *Adventurers*, 302–3. Echoes of the Cagliostro affair can also be found in *Alkhimist bez maski, ili Otkrytoi obman umovoobrazhatel'nago zlatodelaniia, vziatoi iz sochineniia g. professora Gilboa* (Moscow, 1789); and in A. I. Klushin's 1793 one-act play *The Alchemist* published in *Russkaia komediia i komicheskaia opera XVIII veka*, ed. P. N. Berkov (Moscow-Leningrad, 1950), 465–83.

58. Catherine actually did ridicule Freemasonry as simply a form of deception—around the time of Cagliostro's visit (although no direct reference to him is made)—in her short anonymous pamphlet *The Secret of the Anti-Absurd Society, Discovered by Someone Who Isn't a Member*, published in French, German, and Russian in 1780 and widely advertised in the *St. Petersburg News* and the *Moscow News*. For the text and for useful commentary, see *Sochineniia Imperatritsy Ekateriny II*, vol. 5.

59. This is the notion propounded by Bogoliubov (*Novikov i ego vremia*, 357), albeit without any evidence, as well as by Semeka ("Russkie rozenkreitsery," 397–400), despite the fact that Catherine's letters to Brius from this period make no mention of Novikov or his fellow Moscow Masons (see her letters in *RA*, bk. 1 [1872]: 258–74).

60. Reports on Cagliostro's troubles in Paris can be found, for example, in the *Mirror of the World*, no. 5 (5 March 1786): 102; no. 15 (15 May 1786): 324. The preface to the Moscow edition of Cagliostro's *Treatise* states that the work is being published in order to feed the great demand for information on the Diamond Necklace Affair, which the public had so avidly followed in the *Moscow News*. *Memorial grafa Kalliostro*, "Izvestie." Another possible impetus behind Catherine's plays may have been the Illuminati scandal that erupted in Bavaria between 1784 and 1785 and reverberated throughout Europe. See Roberts, *Mythology*, 118–45.

61. See "Pis'ma Imperatritsy Ekateriny II k Grimmu (1774–1796)," 212–13, 362, 366–67, 373–75, 377–79. Grimm apparently sent a copy of Cagliostro's memoirs to the empress, which she admits to having read (p. 375).

62. This is, in fact, the exact praise Zimmermann accords Catherine after reading her plays, when he writes: "The South no longer enlightens the North, but the North enlightens the South; now enlightenment comes to us from the banks of the Neva." *Filosoficheskaia i politicheskaia perepiska*, 51–52. The complete correspondence concerning the plays and the themes they touch upon can be found in Wilfrid-René Chetteoui, *Cagliostro et Catherine II: La satire impériale contre le mage* (Paris, 1947).

63. Grete de Francesco, *The Power of the Charlatan*, trans. Miriam Beard (New Haven, 1939), 196–204. News of Graham's deeds traveled to Russia where they were publicized in the *Moscow News* in April 1784. See J. T. Alexander, "A Russian Reflection of Dr. James Graham's 'Strange Establishment'," *Newsletter of the Study Group on Eighteenth-Century Russia* 20 (1992): 28–31. On possibly the most famous healer of the century, the Viennese physician Franz Mesmer, see Robert Darnton, *Mesmerism and the End of the Enlightenment in France* (Cambridge, Mass., 1968).

64. The subject has yet to attract historians of Russia. For a few brief remarks, see Karnovich, *Zamechatel'nyia i zagadochnyia lichnosti*, 113.

65. *Moskovskiia vedomosti*, 27 November 1784, no. 95, n.p.

66. Derzhaven, *Sochineniia Derzhavina*, ed. Grot, vol. 3, pt. 3 (St. Petersburg, 1870), 505.

67. Karlinsky, *Russian Drama*, 124–25.

68. Catherine II, "Pis'ma i reskripty imperatritsy Ekateriny II-i k Moskovskim glavnokomanduiushchim," *RA*, bk. 1 (1872): 290. See also her letter of 3 January 1790 on a similar Polish cheat (p. 335).

69. OR RGB, f. 147, No. 90 /M. 1967/, ll. 19–19ob.

70. Ia. L. Barskov, *Perepiska moskovskikh masonov XVIII-go veka, 1780–1792 gg.* (Petrograd, 1915), 241.

71. Barskov, *Perepiska*, 252–55. José de Ribas, a Neopolitan adventurer in Russian service, was active in several St. Petersburg lodges in the early 1780s. Georg Rosenberg, a native of Germany forced to quit his homeland, is best remembered for undertaking a trip to Sweden on behalf of a group of Russian Masons, a trip that led to his expulsion from the order in 1781. A prominent figure in Russian Masonic circles for several decades, F. P. Freze (Frese) was a doctor by training and served in the early 1780s as a member of the College of Medicine and as a director of the St. Petersburg Board of Guardians. In another instance, after joining the New Israel Society during his stay in Avignon in 1788, Admiral S. I. Pleshcheev realized upon his return to Russia that he had been thoroughly deceived by the group. See Longinov, *Novikov*, 290. On the New Israel Society and its links to Russian Masons, see G. V. Vernadskii, *Russkoe masonstvo v tsarstvovanie Ekateriny II* (Petrograd, 1917), 80–83.

72. Marie Daniel Bourrée, baron de Corberon, *Un diplomate français à la cour de Catherine II, 1775–1780: Journal intime*, ed. L. H. Lablande (Paris, 1901), 1:147.

73. "Zapiski A. T. Bolotova," *RS* 10, supplement (1872): 935.

74. [Johann Philipp Wegelin], "Pis'mo neizvestnago litsa o moskovskom masonstve XVIII veka," *RA*, bk. 1 (1874): 1034–36; Barskov, *Perepiska*, 218.

75. *Biograficheskii slovar' professorov i prepodavatelei imperatorskago Moskovskogo universiteta* (Moscow, 1855), 576, 586.

76. Other published attacks include *Mason bez maski, ili Podlinnyia tainstva masonskiia* (St. Petersburg, 1784), a translation of Thomas Wilson's *Le Maçon démasqué, ou le vrai Secret des F. M. mis au jour dans toutes ses parties avec sincérité et sans déguisement* (London, 1751); *Mops bez osheinika i bez tsepi, ili Svobodnoe i tochnoe otkrytie tainstv obshchestva imenuiushchagosia Mopsami* (St. Petersburg, 1784), a translation of one section of G. L. Pérau's *L'ordre des francs-maçons trahi, et le secret de Mopses révélé*, which suggests that homosexuality and licentiousness are common among groups like the Masons. Both books were advertised in the *St. Petersburg News*, 29 November 1784, no. 96, 922; 3 December 1784, no. 97, 933; 14 February 1785, no. 13, 126.

77. [S. S. Eli], *Bratskiia uveshchaniia k nekotorym bratiiam svbdnm. kmnshchkm. Pisany bratom Seddagom* (Moscow, 1784). The Russian appears to have been translated from the German. See *SK*, vol. 3 (Moscow, 1966), no. 8589. The bookseller advertising *Fraternal Admonitions* (*Sanktpeterburgskiia vedomosti*, 22 August 1785, no. 67, 690) was also selling *The Freemasons' Pocket Book* published twice in Russia, first as *Zapisnaia knizhka dlia druzei chelovechestva* (St. Petersburg, n.d. [1781?]) and later as *Karmannaia knizhka dlia V*** K*** i dlia tekh, kotorye i ne prinadlezhat k chislu onykh* (Moscow, 1783). See *SK*, vol. 1, no. 2321; and vol. 2, no. 2852. For a discussion of the book's intended audience, see D. D. Lotareva, "Nekotorye istochnikovedcheskie problemy izucheniia masonskoi knizhnosti v Rossii v kontse XVIII–pervoi polovine XIX v.," in *Mirovospriiatie i samosoznanie russkogo obshchestva (XI–XX vv.): Sbornik statei*, ed. L. N. Pushkarev et al. (Moscow, 1994), 154.

78. [Eli], *Bratskiia uveshchaniia*, 26–27. What this "political secret" might be is not clear. The Russian reads: ". . . ne politicheskaia li taina P. S., byli ikh [the founders' of the order] namereniem?" No clues are provided to shed light on what "P. S." might have signified.

79. Ibid., 36–38. On the final page of the first edition of *Fraternal Admonitions*, four other Masonic titles are listed for sale: (1) *Apologiia, ili Zashchishchenie ordena Vol'nykh kamenshchikov. Pisannaia bratom **** chlenom Shotlandskoi ** lozhi, v P*** (Moscow, 1784). Written by J. A. Starck, who had been active in Russian Masonic circles in the 1760s, *Apologiia* was first published in Germany in 1770. It went through two Russian editions in 1784. (2) *Karmannaia knizhka dlia V*** K**** (Moscow, 1783). (3) *Krata Repoa, ili Posviashchenie v drevnee tainoe obshchestvo egipetskikh zhretsov* (Moscow, 1779) by the German K. F. Köppen, in which was depicted a secret society of Egyptian priests with obvious similarities to the Masonic order. Published three times in Russia during the century, *Krata Repoa* appears to have elicited a strong response among the country's readers. See N. I. Novikov, *N. I. Novikov i ego sovremenniki: Izbrannye sochineniia*, ed. I. V. Malyshev (Moscow, 1961), 190–93, 498–99. (4) *Khimicheskaia psaltyr'*, a hermetic work in the spirit of Paracelsus, published in 1784. Most, though not all, of the copies of four of these titles were confiscated by the Moscow authorities and destroyed in 1786; the copies that survived were still openly sold. See Vernadskii, *Russkoe masonstvo*, 125.

80. *Magazin svobodno-kamen'shchicheskoi*, vol. 1, pts. 1–2 (1784): "Izdatel' k Chitateliu," i–ii. The author of these words may well have been Eli himself. See *SK*, vol. 4 (Moscow, 1966), no. 180. This preface, as well as the censor's permission printed on the book's title page, and the book's large print run (possibly as high as 1,200 copies), all belie Novikov's claim, made after his arrest in 1792 and clearly intended to exculpate himself, that this work was intended only for Freemasons. See Novikov, *Novikov i ego sovremenniki*, 448–50.

81. *Magazin svobodno-kamen'shchicheskoi*, vol. 1, pt. 1:109–11, 118.

82. Ibid., pt. 2:73.

83. Ibid., 98–99.

84. Ibid., 14–15. Italics in original.

85. Masons' use of the political languages of the ancients placed them squarely within the practices of the day. On the manipulation of classical iconography by Russia's eighteenth-century rulers, see Richard S. Wortman, *Scenarios of Power: Myth and Ceremony in Russian Monarchy*, vol. 1 (Princeton, 1995). The locus

classicus on civic humanism in early modern Europe is J. G. A. Pocock, *The Machiavellian Moment: Florentine Political Thought and the Atlantic Republican Tradition* (Princeton, 1975). On the contemporary notion of "citizen," see Pierre Rétat, "Citoyen—Sujet, Civisme," in *Handbuch der politisch-sozialer Grundbegriffe in Frankreich, 1680–1820,* ed. Rolf Reichardt and Eberhard Schmitt, Heft 9 (Munich, 1988), 76–105.

86. *Magazin svobodno-kamen'shchicheskoi,* vol. 1, pt. 2:12.

87. Emin, *Mnimyi mudrets,* 58–59.

88. Ibid., 30.

89. Ibid., 31.

90. Novikov, *Novikov i ego sovremenniki,* 150–51. On this society, see W. Gareth Jones, *Nikolay Novikov, Enlightener of Russia* (Cambridge, 1984), 82–95.

91. *Sanktpeterburgskiia vedomosti,* 21 April 1780, no. 32:302–4.

92. D. I. Fonvizin, *Sobranie sochinenii,* ed. G. P. Makogonenko, vol. 2 (Moscow-Leningrad, 1959), 271–72.

93. Derzhavin, *Sochineniia Derzhavina,* vol. 1 (St. Petersburg, 1895), 38–39. See also the commentary on pp. 215–17. For an English translation and a discussion of the work, see Harold B. Segel, ed. and trans., *The Literature of Eighteenth-Century Russia: A History and Anthology,* vol. 2 (New York, 1967), 262–79.

94. In a letter to Baron Grimm dated 11 January 1780. *RA,* bk. 3 (1878): 62.

95. The complete "Daily Notes" are reprinted in *Sochineniia Imperatritsy Ekateriny II,* 5:186–243. According to one source, a society named "Ignoranti Bambinelli" actually did exist and its members included Catherine, L. A. Naryshkin, A. S. Stroganov, and A. P. Shuvalov. Catherine II, *Sochineniia,* ed. V. K. Bylinin and M. P. Odesskii (Moscow, 1990), 517 n. 2.

96. *Sochineniia Imperatritsy Ekateriny II,* 5:213–14. Italics in original. This section of the "Daily Notes" was not originally published in 1783.

97. *Slovar' Akademii Rossiiskoi,* pt. 4 (St. Petersburg, 1793), s.v. "Martyshka." A genus of Old World Monkeys (native to Africa), Cercopithicus, *martyshki* in Russian, comprises twenty species of monkeys and eight species groups. The Green "martyshka", or Green Monkey, is the Cercopithecus aethiops (common name Savanna monkey). *Pifik* most likely refers to "Pithecia." Native to South America, this species of monkey (common name Sakis) is not related to the Cercopithecus. J. R. Napier and P. H. Napier, *The Natural History of Primates* (London, 1985), 119–20, 128, 139.

98. *Rastushchii vinograd* (July 1786): iii.

99. "Apollonian ravings" is most likely a reference to the Pythagorean philosopher and mystic Apollonius of Tyana (born circa 4 B.C.).

100. *Rastushchii vinograd* (July 1786): iv–vii. The italics are in the original.

101. On the appearance of Saint-Martin's book in Russia, see Vl. Tukalevskii, "Iz istorii filosofskikh napravlenii v russkom obshchestve XVIII veka," *Zhurnal Ministerstva narodnago prosveshcheniia,* n.s., 33, no. 5 (1911): 20–25; Vernadskii, *Russkoe masonstvo,* 162–63. According to Elagin, Saint-Martin's book quickly became almost common reading for Russia's readers. "Zapiska o masonstve I. P. Elagina," *RA,* bk. 1 (1864; reprint, 1866): 587.

102. 19 April 1785, no. 32:423–24. An abbreviated advertisement was published later in the year in no. 86.

103. Longinov, *Novikov,* 243–44.

104. "Zapiska o masonstve," 587–89, 601–2; Vernadskii, *Russkoe masonstvo*, 130, 162–63, 170.

105. H. M. Marcard, *Zimmermann's Verhältnisse mit der Kayserin Catharina II. und mit dem Herrn Weikard* (Bremen, 1803), 134–35. Prince Repnin appears to have earned himself a reputation as a "zealous Martinist" at court. See Charles François Philibert Masson, *Secret Memoirs of the Court of Petersburg* (1802; reprint, New York, 1970), 92. Masson makes other references to Martinism on pp. 172, 175–76, 194. Repnin's reputation even rubbed off on individuals who served under him. See A. V. Khrapovitskii, *Pamiatnye zapiski A. V. Khrapovitskogo*, ed. G. N. Gennadi (1862; reprint, Moscow, 1990), 292.

106. Jones, *Nikolay Novikov*, 208.

107. *Rastushchii vinograd* (March 1786): 6–13. The figure upon whom the article is based is the French scholar and self-appointed prophet of ecumenicity Guillaume Postel (1510–1588). He is not known to have any connection to Martinism, whose true founder was Saint-Martin's teacher, the Spanish mystic Martinez de Pasqually. On Postel, see André Nataf, *The Wordsworth Dictionary of the Occult* (Ware, Hertfordshire, 1994), s.v. "Postel." On Martinism and its links with Freemasonry, see Roberts, *Mythology*, 103–5.

108. Act I, Scene 1. Ellipses in original. *On Errors and Truth* and Martinists are also mocked in Emin's *The Sham Sage*. Pypin, *Russkoe masonstvo*, 279.

109. Tsitsianov, "Pis'ma kniazia P. D. Tsitsianova," 2113. It is not clear whether Tsitsianov knew that Zinov'ev, then in France, had actually met Saint-Martin during a stay in Lyon and had even traveled with the great "philosophe inconnu" to Paris.

110. Ibid., 2119–21. The second work to which Tsitsianov refers is the Swedish philosopher Emanuel Swedenborg's *Arcana coelestia* (1749–1756).

111. For Russia's initial reaction to the revolution from which the above information is drawn, see M. M. Shtrange, *Russkoe obshchestvo i frantsuzskaia revoliutsiia, 1789–1794 gg.* (Moscow, 1956), 47–88.

112. Ibid., 101–3.

113. "Moskovskiia pis'ma v poslednie gody ekaterinskago tsarstvovaniia," *RA*, bk. 3 (1876): 277–78. On the popularity of French revolutionary style, see also D. P. Runich, "Iz zapisok D. P. Runicha," *RS* 110 (January 1901): 53–54.

114. *Sanktpeterburgskiia vedomosti* (1790), no. 16, 247; no. 20, 315. Quoted in Shtrange, *Russkoe obshchestvo*, 49–51.

115. Roberts, *Mythology*, 168–74.

116. Jacob, *Living the Enlightenment*, 23–32.

117. Roberts, *Mythology*, 193.

118. Ibid., 210–11.

119. Johannes Rogalla von Bieberstein, "Die These von der freimaurerischen Verschwörung," in *Freimaurer und Geheimbünde im 18. Jahrhundert in Mitteleuropa*, ed. Helmut Reinalter (Frankfurt a. M., 1983), 85, 93; Roberts, *Mythology*, 210.

120. Shtrange, *Russkoe obshchestvo*, 78, 96–97.

121. Barskov, *Perepiska*, 104. What Trubetskoi, himself a Freemason, appears not to have known is that one of *Bouche de fer*'s directors, Nicholas de Bonneville, had in fact been a Freemason and did advocate a revolutionary position for the lodges. Gary Kates, *The "Cercle Social," the Girondins, and the French Revolution* (Princeton, 1985), 89–92; Roberts, *Mythology*, 159–60. According to Roberts, both of the journal's directors, Bonneville and the abbé Fauchet, had been Freemasons.

122. "Zapiski Aleksandra Mikhailovicha Turgeneva," *RS* 53 (January 1887): 88.

123. Antoine Faivre, "Friedrich Tieman und seine deutschen und russischen Freunde," in *Beförderer der Aufklärung in Mittel- und Osteuropa: Freimaurer, Gesellschaften, Clubs*, ed. Éva H. Balázs et al. (Berlin, 1979), 299.

124. A. N. Radishchev, *A Journey from St. Petersburg to Moscow*, trans. Leo Weiner, ed. and intro. by Roderick Page Thaler (Cambridge, Mass., 1958), 80, 239. In her comments, Catherine actually uses "Martinists" in both ways—that is, as mystical theosophists or as "semi-sophists" like "Rousseau, the Abbé Raynal, and similar hypochondriacs." See pp. 241, 244. Fonvizin also mocked the Martinists' libidinal deficiency in *Drug chestnykh liudei, ili starodum*. In *Sobranie sochinenii*, ed. G. P. Makogonenko, vol. 2 (Moscow-Leningrad, 1959), 62.

125. Barskov, *Perepiska*, 15. For a useful discussion of the intellectual and personal relationship between Kutuzov and Radishchev, see Iu. M. Lotman, "'Sochuvstvennik' A. N. Radishcheva A. M. Kutuzov i ego pis'ma k I. P. Turgenevu," *Uchenye zapiski Tartuskogo universiteta* 139 (1963): 281–334.

126. Barskov, *Perepiska*, 15.

127. Ibid., 22.

128. Ibid., 96–97.

129. Ibid., 200, 203. It should be noted that there is no evidence of any Illuminati presence in Russia in the eighteenth century. The history of the Illuminati is discussed in Richard van Dülmen, *Der Geheimbund der Illuminaten* (Stuttgart, 1975).

130. Barskov, *Perepiska*, ix–xi.

131. The process by which these Masons (Lopukhin, Kutuzov, Trubetskoi, Novikov, and I. P. Turgenev) out of all the brothers active in the movement and out of all the hundreds of enthusiasts for Saint-Martin's *On Errors and Truth* acquired this identity is still not fully understood. One important factor was Radishchev's dedication of his *Journey* to Kutuzov, which, once the scandal surfaced, linked Radishchev—at least in the mind of the authorities—to Kutuzov's Masonic comrades in Moscow.

132. Lopukhin, *Zapiski senatora I. V. Lopukhina* (1859; reprint, Moscow, 1990), 28.

133. Turgenev, "Zapiski A. M. Turgeneva," 88.

134. [P. S. Baturin], *Izsledovanie knigi O zabluzhdeniiakh i istine. Sochineno osoblivym obshchestvom odnogo gubernskago goroda* (Tula, 1790), 63–64, 79, 214–15. Published anonymously by a fictitious provincial society of learned men, the work's author was ascertained only much later. On Baturin, a middle-rank official in the provincial administration and writer of comedies, see *Slovar' russkikh pisatelei XVIII veka*, ed. A. M. Panchenko, vol. 1 (Leningrad, 1988), s.v. "Baturin Pafnutii Sergeevich."

135. Shtrange, *Russkoe obshchestvo*, 111. There was even fear at court in 1792 that a Jacobin club was being organized in St. Petersburg. Khrapovitskii, *Pamiatnye zapiski A. V. Khrapovitskogo*, 278.

136. Lopukhin, *Zapiski Lopukhina*, 29–30. The original Russian version is reproduced on pp. 31–37.

137. Ibid., 38. The Russian version is *Kto mozhet byt' dobrym grazhdaninom i vernym poddannym?* (Moscow, 1796; 1798; n.d.). See *SK*, vol. 3, nos. 7405–7. On I. P. Turgenev, who later served as director of Moscow University in the reign of Paul I,

see E. Tarasov, "K istorii russkago obshchestva vtoroi poloviny XVIII stoletii: Mason I. P. Turgenev," *Zhurnal Ministerstva narodnago prosveshcheniia,* n.s., 51, no. 6 (1914): 129–75.

138. See Roberts, *Mythology.*

139. P. N. Berkov, *Istoriia russkoi zhurnalistiki XVIII veka* (Moscow-Leningrad, 1952), 116.

140. Jones, *Nikolay Novikov,* 206–15.

141. N. I. Novikov, *Izbrannye sochineniia,* ed. G. P. Makogonenko (Moscow-Leningrad, 1951), 606–62, 671–72.

142. Longinov, *Novikov,* 348–52.

143. Ibid., 289, 354–55; Bogoliubov, *Novikov i ego vremia,* 442–44.

144. Longinov, *Novikov,* 304; A. I. Serkov, "Rossiiskoe masonstvo: Entsiklope-dicheskii slovar'," vol. 1, Masony v Rossii (1731–1799)," unpublished manuscript, Moscow, n.d., 846, 1020. According to Serkov, it is possible that even as late as 1797 a lodge was meeting in Petrozavodsk. Pp. 1019–20.

145. Catherine II, "Pis'ma i reskripty," 538–39; Friedrichs, *Geschichte,* 89–91; N. S. Ivanina, "K istorii masonstva v Rossii," *RS* 35 (September 1882): 544.

146. Klaus Epstein, *The Genesis of German Conservatism* (Princeton, 1966), 506–35.

147. Manfred Agethen, *Geheimbund und Utopie: Illuminaten, Freimaurer und deutsche Spätaufklärung* (Munich, 1984), 63–64; Robert Freke Gould, *The History of Freemasonry,* vol. 4 (New York, 1889), 103, 108; Jacob, *Living the Enlightenment,* 172, 176, 210–11; Kates, "Cercle Social," 91; Helmut Reinalter, *Aufgeklärter Absolutismus und Revolution: Zur Geschichte des Jakobinertums und der frühdemokratischen Bestrebun-gen in der Habsburgmonarchie* (Vienna, 1980), 186–218; Roberts, *Mythology,* 221; R. William Weisberger, *Speculative Freemasonry and the Enlightenment: A Study of the Craft in London, Paris, Prague, and Vienna,* East European Monographs, 367 (New York, 1993), 105–7.

148. Longinov, *Novikov,* 362–64.

149. Roderick E. McGrew, *Paul I of Russia, 1754–1801* (New York, 1992), 195–96.

150. Friedrichs, *Geschichte,* 91–98; E. S. Shumigorskii, "Imperator Pavel I i ma-sonstvo," in *Masonstvo v ego proshlom i nastoiashchem,* ed. S. P. Mel'gunov and N. P. Sidorov (1914–1915; reprint, Moscow, 1991), 2:135–52. Serkov lists Paul as probably having been initiated into the Masonic order in August 1776 during a trip to Ger-many. "Rossiiskoe masonstvo," 557–58, 1092. Paul was rumored to have been mur-dered by the Masons for having broken his Masonic vows and for having failed to patronize the movement. See Shumigorskii, "Imperator Pavel I," in Mel'gunov and Sidorov, *Masonstvo v ego proshlom,* 152.

CONCLUSION

1. The snuffbox is now in the collection of the Hillwood Museum in Washing-ton, D.C. For a photograph and brief description, see Katrina V. H. Taylor, *Russian Art at Hillwood* (Washington, D.C., 1988), 38–39.

2. See the discussion in Anthony J. La Vopa, "Conceiving a Public: Ideas and Society in Eighteenth-Century Europe," *Journal of Modern History* 64 (March 1992): 79–116.

3. *Nationalism: Five Roads to Modernity* (Cambridge, Mass., 1992), 488.

4. "Transfiguration and Modernization: The Paradoxes of Social Disciplining, Paedagogical Leadership, and the Enlightenment in Eighteenth-Century Russia," in *Alteuropa—Ancien Régime—Frühe Neuzeit: Probleme und Methoden der Forschung*, ed. Hans Erich Bödeker and Ernst Hinrichs (Stuttgart, 1991), 109.

5. See Charles Taylor, "Invoking Civil Society," chap. 11 in *Philosophical Arguments* (Cambridge, Mass., 1995).

6. Elise Kimerling Wirtschafter, *Social Identity in Imperial Russia* (DeKalb, Ill., 1997), 97–99.

7. Adam B. Seligman, *The Idea of Civil Society* (New York, 1992), 5.

8. Anthony Graham Netting, "Russian Liberalism: The Years of Promise, 1842–1855" (Ph.D. diss., Columbia University, 1967).

9. Tira Sokolovskaia, "Rannee Aleksandrovskoe masonstvo," in *Masonstvo v ego proshlom i nastoiashchem*, ed. S. P. Mel'gunov and N. P. Sidorov (1914–1915; reprint, Moscow, 1991), 2:153–63.

10. O. F. Solov'ev, *Russkoe masonstvo, 1730–1917* (Moscow, 1993), 121.

11. On the Masons, see Tatiana Bakounine, *Le répertoire biographique des francs-maçons russes (XVIIIe et XIXe siècles)*, Collection historique de l'Institut d'Etudes Slaves, 19 (Paris, 1967); on the lodges, A. N. Pypin, *Russkoe masonstvo. XVIII i pervaia chetvert' XIX v.* (Petrograd, 1916), 522–32.

12. Bakounine, *Répertoire biographique*, 42; Pypin, *Russkoe masonstvo*, 380–84; Sokolovskaia, "Ranne Aleksandrovskoe masonstvo," in Mel'gunov and Sidorov, *Masonstvo v ego proshlom*, 176–80.

13. Janet M. Hartley, *Alexander I* (London, 1994), 149–54.

14. "Unichtozhenie masonskikh lozh v Rossi 1822 g." *RS* 18 (March 1877): 463; (April 1877): 641–44, 650.

15. "Unichtozhenie masonskikh lozh," *RS* (April 1877): 650–52.

16. M. I. Pyliaev, *Staraia Moskva: Rasskazy iz byloi zhizni pervoprestol'noi stolitsy* (1891; reprint, Moscow, 1990), 82; Solov'ev, *Russkoe masonstvo*, 131–32. The complex relationship between Decembrism and Freemasonry has long attracted the attention of students of Russia. Solov'ev (pp. 110–25) provides a useful overview of the subject, although his attempt to view the two as completely separate historical developments should be read with caution.

17. Barbara T. Norton, "Russian Political Masonry and the February Revolution of 1917," *International Review of Social History* 28, pt. 2 (1983): 240–58; Nathan Smith, "Political Freemasonry in Russia, 1906–1918: A Discussion of the Sources," *Russian Review* 44, no. 2 (April 1985): 157–73; Solov'ev, *Russkoe masonstvo*, 126–262.

18. Viktor Brachev, "Tainye masonskie obshchestva v SSSR," *Molodaia gvardiia* 3 (1994): 140–58; A. I. Nemirovskii and V. I. Ukolova, *Svet zvezd, ili Poslednii russkii rozenkreitser* (Moscow, 1994); A. I. Serkov, "Pridet li k nam 'Velikii vostok'?" *Rodina* 11 (1993): 85–87. The scanty documentary evidence together with the common attempts at outright obfuscation and fabrication surrounding Freemasonry during the Soviet era—most visible in Brachev's piece—warrant highlighting. On the present revival, see Serkov, "Pridet li k nam 'Velikii vostok'?"; Oleg Zolotov, "Russian Masons True to their Homeland," *Moscow News*, no. 8 (25 February–3 March 1994): 15. Freemasonry's return has been shadowed by the rebirth of an openly expressed fear

of *Zhido-masonstvo* ("Jewmasonry"). See Walter Laqueur, "The Appearance of the *Protocols* and the Great Masonic Plot," chap. 3 in *Black Hundred: The Rise of the Extreme Right in Russia* (New York, 1993). A popular example of recent anti-Masonry is O. A. Platonov, *Trenovyi venets Rossii: Istoriia masonstva, 1731–1995* (Moscow, 1995), which blames the collapse of the Soviet Union on western Freemasons, the CIA, the Rotarians, the Trilateral Commission, and a fifth column within the former Soviet elite.

Selected Bibliography

PRIMARY WORKS

Manuscript Sources

MOSCOW

Center for the Preservation of Historico-Documentary Collections (TsKhIDK)
    fond 175, Masonic Lodges and Chapters for 1781–1939
    fond 1311, Knightly Orders for 1785–1931
    fond 1412, Documentary Materials of Masonic Lodges
Division of Written Sources, State Historical Museum (OPI GIM)
    fond 17, Uvarovs' Personal Collection
    fond 281, Document Collection on the History of Culture, Science, and Social
        Movements
    fond 282, Document Collection of the Museum of the Revolution
    fond 398, P. P. Beketov Collection
    fond 440, I. E. Zabelin Collection
    fond 450, E. V. Barsov Collection
Manuscript Division, Russian State Library (OR RGB)
    fond 14, V. S. Arsen'ev Collection of Masonic Manuscripts
    fond 147, S. S. Lanskoi and S. V. Eshevskii Collection of Masonic Manuscripts
    fond 178, Museum Collection
    fond 237, D. I. Popov Collection
Russian State Archive of Early Acts (RGADA)
    fond 8, Kalinkin House *(Kalinkin dom)* and Files on Crimes against Morality
    fond 10, Private Office of Catherine II
    fond 17, Science, Literature, Art
    fond 168, Relations of Russian Sovereigns with Governmental Posts and with
        Officials
Russian State Archive of Literature and Art (RGALI)
    fond 191, Efremov
    fond 442, M. K. and T. O. Sokolovskii

fond 1189, M. M. Kheraskov
fond 1270, N. I. Novikov
fond 1764, I. P. Elagin

St. Petersburg

Central Scientific Library—Rare Book Room, State Hermitage (TsNB-KRK GE)
No. 150554
Division of Manuscripts and Rare Books, Russian Academy of Sciences Library
(ORRK BAN)
file 1.5.41, Manuscript Collection of G. I. Kleinikov
file 17.16.37, Manuscript Collection of A. A. Kunik
Manuscript Division, Institute of Russian Literature (Pushkin House), Russian
Academy of Sciences, (IRLI, PD)
fond 309, Turgenev Collection
Manuscript Division, Russian National Library (OR RNB)
fond 487, N. M. Mikhailovskii Collection
fond 550, Principal Collection of the Manuscript Book (OSRK)
Titovskoe Collection, No. 4419

Printed Sources

*Alkhimist bez maski, ili Otkrytoi obman umovoobrazhatel'nago zlatodelaniia, vziatoi iz
sochinenii g. professora Gilboa.* Moscow: Universitetskaia Tipografiia, 1789.
Arapov, Pimen, comp. *Letopis' russkago teatra.* St. Petersburg: Tipografiia N. Tiblena,
1861.
Barkov, Ivan. *Devich'ia igrushka, ili Sochineniia gospodina Barkova.* Edited by A. Zorin
and N. Sapov. Moscow: Ladomir, 1992.
Barskov, Ia. L. *Perepiska moskovskikh masonov XVIII veka, 1780–1792 gg.* Petrograd: Imperatorskaia Akademiia nauk, 1915.
[Baturin, P. S.]. *Izsledovanie knigi O zabliuzhdeniiakh i istine. Sochineno osoblivym obshchestvom odnogo gubernskago goroda.* Tula: N.p., 1790.
[Bazhanov, Georgii, trans.]. *Karmannaia knizhka chestnago cheloveka, ili Nuzhnyia prav-
ila vo vsiakom meste i vo vsiakoe vremia.* St. Petersburg: [Tipografiia V. Plav-
il'shchikova], 1794.
Bergholz, Friedrich Wilhelm von. *Dnevnik kamer"-iunkera F. V. Berkhgol'tsa,
1721–1725.* Moscow: Universitetskaia Tipografiia, 1902–1903.
Böber, Johann. *Auswahl von Freymaurerliedern: Durch die E. Loge Muse Urania gesam-
melt.* N.p., 1788.
Bokova, V. M., and N. I. Tsimbaev, eds. *Zolotoi vek Ekateriny Velikoi: Vospominaniia.*
Moscow: Izdatel'stvo MGU, 1996.
Bolotov, A. T. "Zapiski A. T. Bolotova." *Russkaia starina* 10, supplement (1872):
897–991; 12, supplement (1872): 1122–244.
Borowski, Ludwig Ernst. *Cagliostro: Einer der merkwürdigsten adventheurer unsres
jahrhunderts.* 2nd ed. Königsberg: G. L. Hartung, 1790.
Bourrée, Marie Daniel, baron de Corberon. *Un diplomate français à la cour de Catherine*

II, 1775–1780: Journal intime. 2 vols. Edited by L. H. Lablande. Paris: Plon-Nourrit, 1901.

Cagliostro, Count Alessandro di [Giuseppe Balsamo]. *Confessions du comte de C*** avec l'histoire de ses voyages en Russie, Turquie, Italie, et dans les pyramides d'E-gypte.* Au Caire [Paris]: Cailleau, 1787.

———. *Memorial grafa Kalliostro protiv gospodina general prokurora obviniaiushchago ego, pisannoi im samim.* Moscow: Tipografiia Ponomareva, 1786.

———. *Opravdanie grafa de Kalliostro po delu kardinala Rogana o pokupke slavnago sklavazha vo Frantsii.* St. Petersburg: Shnor, 1786.

Catherine II. *Documents of Catherine the Great: The Correspondence with Voltaire and the "Instruction" of l767 in the English Text of 1768.* Edited by W. F. Reddaway. Cambridge: Cambridge University Press, 1931.

———. *Filosoficheskaia i politicheskaia perepiska Imperatritsy Ekateriny II s Doktorom Tsimmermanom, s 1785 po 1792 god.* St. Petersburg: Imperatorskaia Tipografiia, 1803.

———. "Pis'ma Ekateriny Vtoroi k baronu Grimmu." *Russkii arkhiv,* bk. 3 (1878): 5–240.

———. "Pis'ma i reskripty imperatritsy Ekateriny II-i k Moskovskim glavnokomanduiushchim." *Russkii arkhiv,* bk. 1 (1872): 225–336, 533–80, 865–80.

———. "Pis'ma Imperatritsy Ekateriny II k Grimmu (1774–1796)." *Sbornik Imperatorskago russkago istoricheskago obshchestva.* Vol. 23. St. Petersburg, 1878.

———. *Sochineniia.* Edited by V. K. Bylinin and M. P. Odesskii. Moscow: Sovremennik, 1990.

———. *Sochineniia Imperatritsy Ekateriny II.* 12 vols. Edited by A. N. Pypin. St. Petersburg: Imperatorskaia Akademiia nauk, 1901–1907.

———. *Taina protivo-nelepago obshchestva, (Anti-absurde) otkrytaia ne prichastnym onomu.* [St. Petersburg: Tipografiia Veitbrekhta i Shnora], 1759 [1780].

Custine, Marquis de. *Empire of the Czar: A Journey through Eternal Russia.* New York: Doubleday, 1989.

Dal', Vladimir. *Tolkovyi slovar' zhivago velikorusskago iazyka.* Vol 3. St. Petersburg-Moscow: Izdatel'stvo Tovarishchestva M. O. Volfa, 1882.

Derzhavin, G. R. *Sochineniia.* 9 vols. Edited by Ia. Grot. St. Petersburg: Akademiia nauk, 1864–1883.

———. *Sochineniia Derzhavina.* Vol. 1. St. Petersburg: N. F. Mertz, 1895.

Dobrynin, Gavriil. *Istinnoe povestvovanie, ili zhizn' Gavriila Dobrynina, im samim napisannaia v Mogileve i v Vitebske, 1752–1823.* St. Petersburg: V. I. Golovin, 1872.

[Doillot]. *Vozrazhenie so storony grafiny de Valua–la Mott, na opravdanie grafa de Kalliostro.* St. Petersburg: Shnor, 1786.

Dolzhnosti chestnago cheloveka. St. Petersburg: N.p., 1798.

"Donesenie o masonakh." *Letopisi russkoi literatury i drevnosti* 4, sect. 3 (1862): 49–52.

Druzheskie sovety molodomu cheloveku, nachinaiushchemu zhit' v svete. Moscow: Kaiserliche Universitäts-Buchdruckerei, 1762.

Druzheskoe Uchenoe Obshchestvo s dostodolzhnym vysokopochitaniem priglashaet sim i prosit imeniteishikh liubitelei nauk i pokrovitelei uchenosti udostoit svoim prisutstviem torzhestvennoe ego otkrytie, imeiushchee byt' v dome ego vysokorodiia, Petra Alekseevicha Tatishcheva, noiabria 6 dnia l782 goda. Moscow: Universitetskaia Tipografiia, 1782.

Elagin, I. P. "Zapiska o masonstve I. P. Elagina." *Russkii arkhiv,* bk. 1 (1864; reprinted, 1866): 586–604.

[Eli, S. S.]. *Bratskiia uveshchaniia k nekotorym bratiiam svbdnm. kmnshchkm. Pisany bratom Seddagom.* Moscow: Tipografiia N. Razskazova, 1784.

Emin, N. F. *Mnimyi mudrets: Komediia v piati deistviiakh.* St. Petersburg: Imperatorskaia Akademiia nauk, 1786.

Fonvizin, D. I. *Dramatic Works of D. I. Fonvizin.* Edited and translated by Marvin Kantor. Frankfurt a. M.: Peter Lang, 1974.

————. *Sobranie sochinenii.* Vol. 2. Edited by G. P. Makogonenko. Moscow-Leningrad: Gosudarstvennoe izdatel'stvo khudozhestvennoi literatury, 1959.

Freimäurerlieder, zum Gebrauch der E. Loge Muse Urania. St. Petersburg: Schnoor, 1783.

[Göhring, Joseph Friedrich]. *Dolzhnosti brat'ev Z. R. K. drevniia sistemy, govorennyia Khrizoferonom, v sobraniiakh iunioratskikh, s prisovokupleniem nekotorykh drugikh rechei drugikh brat'ev.* Moscow: Tipografiia I. Lopukhina, 1784.

Grot, J. C. *Uchrezhdenie osnovannago v Sanktpeterburge na smertnye sluchai obshchestva.* St. Petersburg: Veitbrekht i Shnor, 1780.

Heyking, Baron Karl-Heinrich. "Vospominaniia senatora barona Karla Geikinga." *Russkaia starina* 91 (August 1897): 291–308; (September 1897): 517–37.

Ivanina, N. S., ed. and trans. "K istorii masonstva v Rossii." Pts. 1 and 2. *Russkaia starina* 35 (September 1882): 533–60; 36 (October 1882): 61–76.

Iz bumag mitropolita moskovskago Platona. Moscow: Imperatorskoe Obshchestvo istorii i drevnostei rossiiskikh pri Moskovskom universitete, 1882.

Karamzin, N. M. *Izbrannye sochineniia.* Vol. 2. Edited by G. P. Makogonenko. Moscow-Leningrad: Khudozhestvennaia literatura, 1964.

Karmannaia knizhka, dlia razmyshliaiushchikh iunoshei, sluzhashchaia k priiatnomu i poleznomu ikh uprazhneniiu. Moscow: Gubernskaia Tipografiia, 1800.

Karmannaia knizhka dlia V*** K*** i dlia tekh, kotorye i ne prinadlezhat k chislu onykh. Moscow: Universitetskaia Tipografiia, 1783.

Khrapovitskii, A. V. *Pamiatnye zapiski A. V. Khrapovitskogo.* Edited by G. N. Gennadi. 1862. Reprint, Moscow: V/O Soiuzteatr STD SSSR, 1990.

Klushin, A. I. *Alkhimist.* In *Russkaia komediia i komicheskaia opera XVIII veka,* edited by P. N. Berkov. Moscow-Leningrad: Iskusstvo, 1950.

Kniazhnin, Ia. B. *Izbrannoe.* Edited by A. P. Valagin. Moscow: Pravda, 1991.

————. *Izbrannye proizvedeniia.* Edited by L. I. Kulakova. Leningrad: Sovetskii pisatel', 1961.

[Köppen, K. F.]. *Krata Repoa, ili Posviashchenie v drevnee tainoe obshchestvo egipetskikh zhretsov.* Moscow: [Tipografiia N. Novikova], 1779.

Kurganov, N. G. *Rossiiskaia universal'naia grammatika, ili Vseobshchee pismoslovie, predlagaiushchee legchaishii sposob osnovatel'nago ucheniia ruskomu [sic] iazyku s sedm'iu prisovokupleniiami raznykh uchebnykh i poleznozabavnykh veshchei.* St. Petersburg: [Tipografiia Morskago kadetskago korpusa], 1769.

Lopukhin, I. V. *Zapiski senatora I. V. Lopukhina.* 1859. Reprint, Moscow: Akademiia nauk SSSR, 1990.

Masson, Charles François Philibert. *Secret Memoirs of the Court of Petersburg.* 1802. Reprint, New York: Arno Press and the New York Times, 1970.

Moshchinskii, A. F. [Moszynski, A. F.]. *Kalliostr poznannyi v Varshave, ili Dostovernoe opisanie khimicheskikh i magicheskikh ego deistvii, proizvodimikh v sem stolichnom gorode v 1780*. Moscow: Senatskaia Tipografiia, 1788.

"Moskovskiia pis'ma v poslednie gody ekaterinskago tsarstvovaniia." *Russkii arkhiv*, bk. 3 (1876): 257–84.

Nordstet, Ivan. *Rossiiskii, s nemetskim i frantsuzskim perevodami, slovar', sochinennyi nadvornym sovetnikom Ivanom Nordstetom*. 2 Pts. St. Petersburg: I. K. Shnor, 1780–1782.

Novikov, N. I. *Izbrannye sochineniia*. Edited by G. P. Makogonenko. Moscow-Leningrad: Gosudarstvennoe izdatel'stvo khudozhestvennoi literatury, 1951.

———. *N. I. Novikov i ego sovremenniki: Izbrannye sochineniia*. Edited by I. V. Malyshev. Moscow: Akademiia nauk SSSR, 1961.

———. *Satiricheskie zhurnaly N. I. Novikova*. Edited by P. N. Berkov. Moscow-Leningrad: Akademiia nauk SSSR, 1951.

O dolzhnostiakh cheloveka i grazhdanina, kniga k chteniiu opredelennaia v narodnykh gorodskikh uchilishchakh Rossiiskoi Imperii, izdannaia po Vysochaishemu Poveleniiu Tsarstvuiushchei Imperatritsy Ekateriny Vtoryia. St. Petersburg: [Tipografiia Akademii nauk], 1783.

[Pérau, G. L.]. *Mops bez osheinika i bez tsepi, ili Svobodnoe i tochnoe otkrytie tainstv obshchestva imenuiushchagosia Mopsami*. St. Petersburg: Khristofor Genning, 1784.

"Plan publichnoi biblioteki v S.-Peterburge, sostavlennyi v 1766 g." *Bibliograficheskiia zapiski* 3, no. 3 (1861): 70–80.

Polnoe sobranie zakonov Rossiiskoi imperii. 1st series. 46 vols. St. Petersburg: Tipografiia II otdeleniia sobstvennoi Ego Imperatorskago Velichestva kantseliarii, 1830.

"Psal'ma, na oblichenie frank-masonov." *Bibliograficheskiia zapiski* 3, no. 3 (1861): 70.

Radishchev, A. N. *A Journey from St. Petersburg to Moscow*. Translated by Leo Weiner. Edited and with introduction by Roderick Page Thaler. Cambridge, Mass.: Harvard University Press, 1958.

———. *Polnoe sobranie sochinenii*. Vol. 2. Edited by V. V. Kallash. Moscow: V. M. Sablin, 1907.

Recke, Charlotte von der. *Opisanie prebyvaniia v Mitave izvestnago Kaliostra na 1779 god, i proizvedennykh im tamo magicheskikh deistvii*. St. Petersburg: Shnor, 1787.

Reinbeck, Georg. *Travels from St. Petersburgh through Moscow, Grodno, Warsaw, Breslaw, etc. to Germany, in the Year 1805*. Vol. 6 of *A Collection of Modern and Contemporary Voyages and Travels*. Edited by Richard Phillips. London: J. G. Barnard, 1807.

Repnin, N. V. *Les Fruits de la grace* [sic], *ou Opuscules spirituels de deux amateurs de la Sagesse*. N.p., 1790.

Richter, Johann. *Moskva–nachertanie, perevod s nemetskago*. St. Petersburg: Akademiia nauk, 1801.

Robison, John. *Proofs of a Conspiracy against all the Religions and Governments of Europe, Carried on in the Secret Meetings of Free Masons, Illuminati, and Reading Societies*. 3rd ed. London: Printed for T. Cadell, jun. and W. Davies, Strand; and W. Creech, Edinburgh, 1798.

Runich, D. P. "Iz zapisok D. P. Runicha." *Russkaia starina* 110 (January 1901): 47–77.

Saint-Martin, Louis Claude de. *O zabliuzhdeniiakh i istine, ili Vozzvanie chelovecheskago roda k vseobshchemu nachalu znaniia*. Moscow: Vol'naia Tipografiia I. Lopukhina, 1785.

Savva, V. I. "Iz dnevnika masona 1775–1776 gg." *Chteniia v Imperatorskom Ob-*

shchestve istorii i drevnostei rossiiskikh 4 (1908): 1–15.

Segel, Harold B., ed. and trans. *The Literature of Eighteenth-Century Russia: A History and Anthology*. 2 vols. New York: E. P. Dutton, 1967.

Shcherbatov, Prince M. M. *On the Corruption of Morals in Russia*. Edited and translated by A. Lentin. Cambridge: Cambridge University Press, 1969.

Slovar' Akademii Rossiiskoi. 6 vols. St. Petersburg: Imperatorskaia Akademiia nauk, 1789–1794.

[Starck, J. A.]. *Apologiia, ili Zashchishchenie ordena Vol'nykh kamenshchikov. Pisannaia bratom **** chlenom Shotlandskoi ** lozhi, v P***. Moscow: Tipografiia N. Razskazova, 1784.

Stennik, Iu. V., ed. *Russkaia satiricheskaia proza XVIII veka*. Leningrad: Izdatel'stvo Leningradskogo universiteta, 1986.

Sumarokov, A. P. *Polnoe sobranie vsekh sochinenii*. 2nd ed. 10 vols. Edited by N. I. Novikov. Moscow: Universitetskaia Tipografiia, 1787.

Tooke, W. *History of Russia: From the Foundation of the Monarchy by Riurik to the Accession of Catherine the Second*. 2 vols. London: Longman, 1800.

Tsitsianov, P. D. "Pis'ma kniazia Pavla Dmitrievicha Tsitsianova k Vasiliiu Nikolaevichu Zinov'evu." *Russkii arkhiv*, bk. 2 (1872): 2100–173.

Turgenev, A. M. "Zapiski Aleksandra Mikhailovicha Turgeneva." *Russkaia starina* 53 (January 1887): 77–106.

Turgenev, I. P. *Kto mozhet byt' dobrym grazhdaninom i vernym poddannym?* Moscow: Tipografiia Ridigera i Klaudiia, 1796, 1798, n.d.

"Unichtozhenie masonskikh lozh v Rossii 1822 g." *Russkaia starina* 18 (March 1877): 455–79; (April 1877): 641–64.

"Ustav vol'nykh kamenshchikov." *More* 25–26 (1907): 769–80.

Weber, Christian. "Zapiski Vebera o Petre Velikom i ego preobrazovaniiakh," *Russkii arkhiv*, bk. 1 (1872): 1057–168; bk. 2 (1872): 1334–457, 1613–704.

[Wegelin, Johann Philipp]. "Pis'mo neizvestnago litsa o Moskovskom masonstve XVIII veka." *Russkii arkhiv*, bk. 1 (1874): 1031–42.

[Wilson, Thomas]. *Mason bez maski, ili Podlinnyia tainstva masonskiia, izdannyia so mnogimi podrobnostiami tochno i bezpristrastno*. St. Petersburg: Khristofor Genning, 1784.

Zakony ustanovlennago v 1792 goda sentiabria v 1 den' v gubernskom gorode Revele Kluba Soglasiia. St. Petersburg: N.p., 1792.

Zapisnaia knizhka dlia druzei chelovechestva. [St. Petersburg: Tipografiia Gosudarstvennoi voennoi kollegii, n.d. (1781?)].

Zinov'ev, V. N. "Zhurnal puteshestviia V. N. Zinov'eva po Germanii, Italii, Frantsii i Anglii v 1784–1788 gg." *Russkaia starina* 23 (October 1878): 207–40; (November 1878): 399–440; (December 1878): 593–630.

Newspapers and Journals

Ezhemesiachnyia sochineniia k pol'ze i uveseleniiu sluzhashchiia (1755–1764) (Monthly Compositions for Profit and Amusement)

Magazin svobodno-kamen'shchicheskoi (1784) (Free-Masonic Magazine)

Moskovskiia vedomosti (1756–1800) (Moscow News)

Moskovskoe ezhemesiachnoe izdanie (1781) (Moscow Monthly Edition)

Novyia ezhemesiachnyia sochineniia (1786–1796) (New Monthly Compositions)
Prazdnoe vremia v pol'zu upotreblennoe (1759–1760) (Idle Time for Good Use)
Rastushchii vinograd (1785–1787) (Growing Vine)
Sankt Peterburgskii Merkurii (1793–1794) (St. Petersburg Mercury)
Sanktpeterburgskiia vedomosti (1728–1800) (St. Petersburg News)
Sobesednik liubitelei rossiiskago slova (1783–1784) (Companion of Lovers of the Russian Word)
Utrennii svet (1777–1780) (Morning Light)
Vechera (1772) (Evenings)
Vecherniaia zaria (1782) (Evening Glow)
Zerkalo sveta (1786–1787) (Mirror of the World)

SECONDARY WORKS

Abramov, K. I. *Istoriia bibliotechnogo dela v SSSR*. Moscow: Kniga, 1970.

Agethen, Manfred. *Geheimbund und Utopie: Illuminaten, Freimaurer und deutsche Spätaufklärung*. Munich: R. Oldenbourg, 1984.

Alexander, John T. *Catherine the Great: Life and Legend*. New York: Oxford University Press, 1989.

———. "A Russian Reflection of Dr. James Graham's 'Strange Establishment'." *Newsletter of the Study Group on Eighteenth-Century Russia* 20 (1992): 28–31.

Andrew, Joe. *Women in Russian Literature, 1780–1863*. New York: St. Martin's Press, 1988.

Artem'ev, A. I. "Kazanskiia gimnazii v XVIII stoletii." *Zhurnal Ministerstva narodnago prosveshcheniia* 173 (May 1874): 32–98.

Aseev, B. N. *Russkii dramaticheskii teatr ot ego istokov do kontsa XVIII veka*. Moscow: Iskusstvo, 1977.

Augustine, Wilson R. "Notes toward a Portrait of the Eighteenth-Century Russian Nobility." *Canadian Slavic Studies* 4, no. 3 (fall 1970): 373–425.

B., N. "Znachenie fran-masonstva [sic] dlia flota." *More* 11–12 (March 1907): 312–27.

Baehr, Stephen L. *The Paradise Myth in 18th-Century Russia: Utopian Patterns in Early Secular Russian Literature and Culture*. Stanford: Stanford University Press, 1991.

Bakounine, Tatiana. *Le répertoire biographique des franc-maçons russes (XVIIIe et XIXe siècles)*. Collection historique de l'Institut d'Etudes Slaves, 19. Paris: Institut d'études slaves de l'Université de Paris, 1967.

Balázs, Éva H., Ludwig Hammermayer, Hans Wagner, and Jerzy Wojtowicz, ed. *Beförderer der Aufklärung in Mittel- und Osteuropa: Freimaurer, Gesellschaften, Clubs*. Berlin: Ulrich Camen, 1979.

Barenbaum, I. E. *Istoriia knigi*. 2nd ed. Moscow: Kniga, 1984.

Barker-Benfield, G. J. *The Culture of Sensibility: Sex and Society in Eighteenth-Century Britain*. Chicago: University of Chicago Press, 1992.

Bartlett, R. P. "Culture and Enlightenment: Julius von Canitz and the Kazan' *Gimnazii* in the Eighteenth Century." *Canadian-American Slavic Studies* 14, no. 3 (fall 1980): 339–60.

Becker, Marvin B. *The Emergence of Civil Society in the Eighteenth Century: A Privileged Moment in the History of England, Scotland, and France*. Bloomington: Indiana University Press, 1994.

Berkov, P. N. *Istoriia russkoi zhurnalistiki XVIII veka.* Moscow-Leningrad: Izdatel'stvo Akademii nauk, 1952.

Berlin, Isaiah. *The Magus of the North: J. G. Hamann and the Origins of Modern Irrationalism.* New York: Farrar, Straus and Giroux, 1993.

Biograficheskii slovar' professorov i prepodavatelei imperatorskago Moskovskogo universiteta. Moscow: Universitetskaia Tipografiia, 1855.

Black, J. L. *Citizens for the Fatherland: Education, Educators, and Pedagogical Ideals in Eighteenth-Century Russia.* East European Monographs, 53. New York: Columbia University Press, 1979.

Bogoliubov, V. *N. I. Novikov i ego vremia.* Moscow: Izdatel'stvo M. i S. Sabashnikovykh, 1916.

Brachev, Viktor. "Tainye masonskie obshchestva v SSSR." *Molodaia gvardiia* 3 (1994): 140–58.

Brewer, John. *The Pleasures of the Imagination: English Culture in the Eighteenth Century.* New York: Farrar, Straus and Giroux, 1997.

Brown, William Edward. *A History of 18th-Century Russian Literature.* Ann Arbor, Mich.: Ardis, 1980.

Bullock, Steven C. *Revolutionary Brotherhood: Freemasonry and the Transformation of the American Social Order, 1730–1840.* Chapel Hill: University of North Carolina Press, 1996.

Burgess, Malcolm. "Russian Public Theater Audiences of the Eighteenth and Early Nineteenth Centuries." *Slavonic and East European Review* 37, no. 88 (December 1958): 160–83.

Burke, Janet M. "Freemasonry, Friendship and Noblewomen: The Role of the Secret Society in Bringing Enlightenment to Pre-Revolutionary Women Elites." *History of European Ideas* 10, no. 3 (1989): 283–93.

———. "French Women Freemasons into the Age of Charity." Paper presented at the annual meeting of the American Historical Association, Seattle, January 1998.

Burke, Janet M., and Margaret C. Jacob. "French Freemasonry, Women, and Feminist Scholarship." *Journal of Modern History* 68 (September 1996): 513–49.

Calhoun, Craig, ed. *Habermas and the Public Sphere.* Cambridge, Mass.: MIT Press, 1992.

Chartier, Roger. "Civilité." In *Handbuch politisch-sozialer Grundbegriffe in Frankreich, 1680–1820,* edited by Rolf Reichardt and Eberhard Schmitt, Heft 4. Munich: R. Oldenbourg, 1986.

———. *The Cultural Origins of the French Revolution.* Translated by Lydia G. Cochrane. Durham, N.C.: Duke University Press, 1991.

Chechulin, N. D. *Russkoe provintsial'noe obshchestvo vo vtoroi polovine XVIII veka.* St. Petersburg: Tipografiia V. S. Babsheva, 1889.

Chernaia, L. A. "Kontseptsiia lichnosti v russkoi literature vtoroi poloviny XVII–pervoi poloviny XVIII v." In *Razvitie Barokko i zarozhdenie Klassitsizma v Rossii XVII-nachala XVIII v.,* edited by A. N. Robinson. Moscow: Nauka, 1989.

Chetteoui, Wilfrid-René. *Cagliostro et Catherine II: La satire impériale contre le mage.* Paris: Les Éditions des Champs-Élysées, 1947.

Chulkov, N. "F. V. Krechetov—zabytyi radikal'nyi publitsist XVIII veka." In *Literaturnoe nasledstvo.* Vols. 9–10. Moscow: Institut literatury Akademii nauk SSSR, 1933.

Clowes, Edith W., Samuel D. Kassow, and James L. West, eds. *Between Tsar and People: Educated Society and the Quest for Identity in Late Imperial Russia*. Princeton: Princeton University Press, 1991.

Cross, A. G. "British Freemasons in Russia during the Reign of Catherine the Great." *Oxford Slavonic Papers*, n.s., 4 (1971): 43–72.

———. "The British in Catherine's Russia: A Preliminary Survey." In *The Eighteenth Century in Russia*, edited by J. G. Garrard. Oxford: Clarendon Press, 1973.

———. *By the Banks of the Neva: Chapters from the Lives and Careers of the British in Eighteenth-Century Russia*. Cambridge: Cambridge University Press, 1997.

Curl, James Stevens. *The Art and Architecture of Freemasonry*. London: B. T. Batsford, 1991.

Dann, Otto. "Eine höfische Gesellschaft als Lesegesellschaft." In *Lesekulturen im 18. Jahrhundert*, edited by Hans Erich Bödeker. Hamburg: Felix Meiner Verlag, 1992.

Darnton, Robert. "George Washington's False Teeth." *New York Review of Books* 44, no. 5 (27 March 1997): 34–38.

———. *Mesmerism and the End of the Enlightenment in France*. Cambridge, Mass.: Harvard University Press, 1968.

Demkov, M. I. *Istoriia russkoi pedagogii*. 2 vols. St. Petersburg: Tipografiia M. M. Stasiulevicha, 1896–1897.

Donnert, Erich. *Politische Ideologie der russischen Gesellschaft zu Beginn der Regierungszeit Katharinas der II*. Berlin: Akademie Verlag, 1976.

Douglas, Mary. *Natural Symbols: Explorations in Cosmology*. New York: Pantheon Books, 1982.

———. *Purity and Danger: An Analysis of the Concepts of Pollution and Taboo*. London: Routledge, 1966.

Dülmen, Richard van. *Der Geheimbund der Illuminaten: Darstellung, Analyse, Dokumentation*. Stuttgart: F. Frommann, 1975.

———. *Die Gesellschaft der Aufklärer: Zur bürgerlichen Emanzipation und aufklärerischen Kultur in Deutschland*. Frankfurt a. M.: Fischer Verlag, 1986.

Dumas, François Ribadeau. *Cagliostro*. Translated by Elisabeth Abbott. New York: Orion Press, 1967.

Eleonskaia, A. S. "Tvorcheskie vzaimosviazi shkol'nogo i pridvornogo teatrov v Rossii." In *P'esy stolichnykh i provintsial'nykh teatrov pervoi poloviny XVIII v.*, edited by O. A. Derzhavina et al. Moscow: Nauka, 1975.

Eliade, Mircea. *The Sacred and the Profane; the Nature of Religion*. New York: Harper and Row, 1959.

Elias, Norbert. *The Civilizing Process*. 2 vols. New York: Pantheon, 1978, 1982.

Entsiklopedicheskii slovar' Brokgauz-Efron. 82 vols. St. Petersburg: Brokgauz-Efron, 1890–1904.

Epstein, Klaus. *The Genesis of German Conservatism*. Princeton: Princeton University Press, 1966.

Eshevskii, S. V. *Sochineniia po russkoi istorii*. Moscow: Izdatel'stvo M. i I. Sabashnikovykh, 1900.

Fedorovskaia, L. A. "'Azbuk musikiiskogo peniia' iz knig Aleksandra Fomina." In *Kniga v Rossii XVI–seredina XIX v.*, edited by A. A. Zaitseva. Leningrad: BAN, 1987.

Fiering, Norman. *Moral Philosophy at Seventeenth-Century Harvard: A Discipline in Transition.* Chapel Hill: University of North Carolina Press, 1981.

France, Peter. "Polish, Police, *Polis.*" In *Politeness and Its Discontents: Problems in French Classical Culture.* Cambridge Studies in French. Cambridge: Cambridge University Press, 1992.

Francesco, Grete de. *The Power of the Charlatan.* Translated by Miriam Beard. New Haven: Yale University Press, 1939.

Freeze, Gregory L. *The Russian Levites: Parish Clergy in the Eighteenth Century.* Cambridge, Mass.: Harvard University Press, 1977.

———. "The *Soslovie* (Estate) Paradigm and Russian Social History." *American Historical Review* 91, no. 1 (February 1986): 11–36.

Friedrichs, Ernst. *Geschichte der einstigen Maurerei in Rußland.* Berlin: Ernst Friedrich Mittler und Sohn, 1904.

Furet, François. *Interpreting the French Revolution.* Translated by Elborg Forster. Cambridge: Cambridge University Press, 1981.

Geertz, Clifford. *Local Knowledge: Further Essays in Interpretive Anthropology.* New York: Basic Books, 1983.

Gennep, Arnold van. *The Rites of Passage.* 1908. Reprint, Chicago: University of Chicago Press, 1960.

Gervaso, Roberto. *Cagliostro: A Biography.* Translated by Cormac Ó Cuilleanáin. London: Victor Gollancz, 1974.

Gleason, Walter J. *Moral Idealists, Bureaucracy, and Catherine the Great.* New Brunswick, N.J.: Rutgers University Press, 1981.

Goffman, Erving. *The Presentation of the Self in Everyday Life.* New York: Doubleday, 1959.

Goodman, Dena. *The Republic of Letters: A Cultural History of the French Enlightenment.* Ithaca, N.Y.: Cornell University Press, 1994.

Gordon, Daniel. *Citizens without Sovereignty: Equality and Sociability in French Thought, 1670–1789.* Princeton: Princeton University Press, 1994.

Gordon, Daniel, David A. Bell, and Sarah Maza. "The Public Sphere in the Eighteenth Century." *French Historical Studies* 17, no. 4 (fall 1992): 882–956.

Gould, Robert Freke. *The History of Freemasonry, its Antiquities, Symbols, Constitutions, Customs, etc., derived from Official Sources throughout the World.* 4 vols. New York: John C. Yorston and Co., [1884?]–1889.

Greenfeld, Liah. *Nationalism: Five Roads to Modernity.* Cambridge, Mass.: Harvard University Press, 1992.

Greenfeld, Liah, and Michel Martin, eds. *Center: Ideas and Institutions.* Chicago: University of Chicago Press, 1988.

Griffiths, D. M. "Catherine II: The Republican Empress." *Jahrbücher für Geschichte Osteuropas,* n.s., 21, Heft 3 (1973): 323–44.

Gross, Irena Grudzinska. *The Scar of Revolution: Custine, Tocqueville, and the Romantic Imagination.* Berkeley–Los Angeles: University of California Press, 1991.

Grot, Ia. "Zametka o pastore Grota." *Sbornik otdeleniia russkago iazyka i slovesnosti Imperatorskoi Akademii nauk* 5 (1868): 289–92.

Gurevich, Liubov'. *Istoriia russkogo teatral'nogo byta.* Vol. 1. Moscow-Leningrad: Iskusstvo, 1939.

Habermas, Jürgen. *The Structural Transformation of the Public Sphere: An Inquiry into a*

Category of Bourgeois Society. Trans. Thomas Burger with the assistance of Frederick Lawrence. Cambridge, Mass.: MIT Press, 1989.

Hammermayer, Ludwig. *Der Wilhelmsbader Freimaurer-Konvent von 1782*. Wolfenbütteler Studien zur Aufklärung. Band 5/2. Heidelberg: Schneider, 1980.

Hans, Nicholas. "The Moscow School of Mathematics and Navigation (1701)." *Slavonic and East European Review* 29, no. 73 (1951): 532–36.

Hardtwig, Wolfgang. "Eliteanspruch und Geheimnis in den Geheimgesellschaften des 18. Jahrhunderts." In *Aufklärung und Geheimgesellschaften: Zur politischen Funktion und Sozialstruktur der Freimaurerlogen im 18. Jahrhundert*, edited by Helmut Reinalter. Munich: R. Oldenbourg, 1989.

Hartley, Janet M. *Alexander I*. London: Longman, 1994.

Haumant, Emile. *La culture française en Russie (1700–1900)*. 2nd ed. Paris: Librairie Hachette, 1913.

Hittle, J. Michael. *The Service City: State and Townsmen in Russia, 1600–1800*. Cambridge, Mass.: Harvard University Press, 1979.

Hobsbawm, Eric, and Terence Ranger, eds. *The Invention of Tradition*. Cambridge: Cambridge University Press, 1983.

Höfer, Anette, and Rolf Reichardt. "Honnête homme, Honnêteté, Honnête gens." In *Handbuch politisch-sozialer Grundbegriffe in Frankreich, 1680–1820*, edited by Rolf Reichardt and Eberhard Schmitt, Heft 7. Munich: R. Oldenbourg, 1986.

Hölscher, Lucian. "Öffentlichkeit." In *Geschichtliche Grundbegriffe: Historisches Lexikon zur politisch-sozialen Sprache in Deutschland*, edited by Otto Brunner, Werner Conze, and Reinhart Koselleck, Band 4. Stuttgart: Klett-Cotta, 1978.

———. *Öffentlichkeit und Geheimnis: Eine begriffsgeschichtliche Untersuchung zur Entstehung der Öffentlichkeit in der frühen Neuzeit*. Stuttgart: Klett-Cotta, 1979.

Hull, Isabel V. *Sexuality, State, and Civil Society in Germany, 1700–1815*. Ithaca, N.Y.: Cornell University Press, 1996.

Im Hof, Ulrich. *Das gesellige Jahrhundert: Gesellschaft und Gesellschaften im Zeitalter der Aufklärung*. Munich: Beck, 1982.

Istoriia Akademii nauk SSSR. Vol. 1. Moscow-Leningrad: Izdatel'stvo Akademii nauk SSSR, 1958.

Istoriia Biblioteki Akademii nauk SSSR, 1714–1964. Pt. 1. Moscow-Leningrad: Nauka, 1964.

Istoriia Moskovskogo Universiteta. Vol. 1. Moscow: Izdatel'stvo Moskovskogo universiteta, 1955.

Istoriia russkoi muzyki. Vol. 3. Moscow: Muzyka, 1985.

Jacob, Margaret C. *Living the Enlightenment: Freemasonry and Politics in Eighteenth-Century Europe*. New York: Oxford University Press, 1991.

———. "The Mental Landscape of the Public Sphere: A European Perspective." *Eighteenth-Century Studies* 28, no. 1 (fall 1994): 95–113.

Jones, W. Gareth. *Nikolay Novikov, Enlightener of Russia*. Cambridge: Cambridge University Press, 1984.

———. "The Polemics of the 1769 Journals: A Reappraisal." *Canadian-American Slavic Studies* 16, nos. 3–4 (fall–winter 1982): 432–43.

Kant, Immanuel. *Kant's Political Writings*. Edited by Hans Reiss. Translated by H. B. Nisbet. Cambridge: Cambridge University Press, 1970.

Karlinsky, Simon. *Russian Drama from Its Beginnings to the Age of Pushkin*. Berkeley–Los Angeles: University of California Press, 1985.

240   *Selected Bibliography*

Karnovich, E. P. *Zamechatel'nyia i zagadochnyia lichnosti XVIII i XIX stoletii.* 2nd ed. St. Petersburg: A. S. Suvorin, 1893.

Kates, Gary. *The "Cercle Social," the Girondins, and the French Revolution.* Princeton: Princeton University Press, 1985.

Keane, John, ed. *Civil Society and the State: New European Perspectives.* London: Verso, 1988.

Keohane, Nannerl O. *Philosophy and the State in France: The Renaissance to the Enlightenment.* Princeton: Princeton University Press, 1980.

Khodnev, A. I. *Istoriia imperatorskago Vol'nago Ekonomicheskago Obshchestva s 1765 do 1865 goda.* St. Petersburg: Obshchestvennaia pol'za, 1865.

Kizevetter, A. A. "Moskovskii Universitet (istoricheskii ocherk)." In *Moskovskii Universitet, 1755–1930: Iubileinyi sbornik,* edited by V. B. El'iashevich, A. A. Kizevetter, and M. M. Novikov. Paris: Sovremennyia zapiski, 1930.

Klein, Lawrence E. "Gender and the Public/Private Distinction in the Eighteenth Century: Some Questions about Evidence and Analytic Procedure." *Eighteenth-Century Studies* 29, no. 1 (fall 1995): 97–109.

———. *Shaftesbury and the Culture of Politeness: Moral Discourse and Cultural Politics in Early Eighteenth-Century England.* Cambridge: Cambridge University Press, 1994.

Kochetkova, N. D. "Ideino-literaturnye pozitsii masonov 80–90-kh godov XVIII v. i N. M. Karamzin." In *XVIII vek.* Vol. 6. Moscow-Leningrad: Nauka, 1964.

Kollmann, Nancy Shields. "The Seclusion of Elite Muscovite Women." *Russian History* 10, pt. 2 (1983): 170–87.

Koselleck, Reinhart. *Critique and Crisis: Enlightenment and the Pathogenesis of Modern Society.* 1959. English translation, Cambridge, Mass.: MIT Press, 1988.

Kosmolinskaia, G. A. "Pamiatniki uchebno-vospitatel'noi literatury vtoroi poloviny XVIII veka v sobranii nauchnoi biblioteki MGU." In *Iz fonda redkikh knig i rukopisei nauchnoi biblioteki Moskovskogo Universiteta.* Moscow: Izdatel'stvo MGU, 1987.

Kulakova, L. I. "M. M. Kheraskov." In *Istoriia russkoi literatury.* Vol. 4. Moscow-Leningrad: Akademiia nauk, 1947.

Laqueur, Walter. *Black Hundred: The Rise of the Extreme Right in Russia.* New York: HarperCollins, 1993.

La Vopa, Anthony J. "Conceiving a Public: Ideas and Society in Eighteenth-Century Europe." *Journal of Modern History* 64 (March 1992): 79–116.

Lazarchuk, R. M. "Iz istorii provintsial'nogo teatra (teatral'naia zhizn' Vologdy 1780-kh gg.)." In *XVIII vek.* Vol. 15. Leningrad: Nauka, 1986.

———. "Iz istorii provintsial'nogo teatra (Vologodskii publichnyi teatr kontsa XVIII–nachala XIX v.)." In *XVIII vek.* Vol. 18. St. Petersburg: Nauka, 1993.

Ledkovsky, Marina, Charlotte Rosenthal, and Mary Zirin, eds. *Dictionary of Russian Women Writers.* Westport, Conn.: Greenwood Press, 1994.

LeDonne, John. P. *Absolutism and Ruling Class: The Formation of the Russian Political Order, 1700–1825.* New York: Oxford University Press, 1991.

———. *Ruling Russia: Politics and Administration in the Age of Absolutism, 1762–1796.* Princeton: Princeton University Press, 1984.

Longinov, M. N. *Novikov i moskovskie martinisty.* Moscow: Tipografiia Gracheva, 1867.

Lotareva, D. D. "Nekotorye istochnikovedcheskie problemy izucheniia masonskoi knizhnosti v Rossii v kontse XVIII–pervoi polovine XIX v." In *Mirovospriiatie i samosoznanie russkogo obshchestva (XI–XX vv.): Sbornik statei,* edited by L. N. Pushkarev et al. Moscow: Institut rossiiskoi istorii RAN, 1994.

Lotman, Iu. M. *Besedy o russkoi kul'ture: Byt i traditsii russkogo dvorianstva (XVIII–nachalo XIX veka).* St. Petersburg: Iskusstvo-SPB, 1994.

———. "'Sochuvstvennik' A. N. Radishcheva A. M. Kutuzov i ego pis'ma k I. P. Turgenevu." *Uchenye zapiski Tartuskogo universiteta* 139 (1963): 281–334.

Lotman, Iu. M., et al. *Theses on the Semiotic Study of Culture (as Applied to Slavic Texts).* Lisse, Netherlands: Peter de Ridder, 1975.

Mackey, Albert G. *An Encyclopedia of Freemasonry.* Philadelphia: Moss and Co., 1874.

Macoy, Robert. *A Dictionary of Freemasonry.* New York: Bell, 1989.

Madariaga, Isabel de. *Russia in the Age of Catherine the Great.* New Haven: Yale University Press, 1981.

Maikov, L. N. *Ocherki iz istorii russkoi literatury XVII i XVIII stoletii.* St. Petersburg: A. S. Suvorin, 1889.

Makogonenko, G. P. *Nikolai Novikov i russkoe prosveshchenie XVIII veka.* Moscow-Leningrad: Gosudarstvennoe izdatel'stvo khudozhestvennoi literatury, 1951.

Malinovskii, A. F. *Obozrenie Moskvy.* Comp. S. R. Dolgova. Moscow: Moskovskii rabochii, 1992.

Marcard, H. M. *Zimmermann's Verhältnisse mit der Kayserin Catharina II. und mit dem Herrn Weikard.* Bremen: Carl Seyffert, 1803.

Marker, Gary. *Publishing, Printing, and the Origins of Intellectual Life in Russia, 1700–1800.* Princeton: Princeton University Press, 1985.

Martens, Wolfgang. *Die Botschaft der Tugend: Die Aufklärung im Spiegel der deutschen moralischen Wochenschriften.* Stuttgart: Metzler, 1968.

Martynov, Aleksei. "Moskovskaia starina: Arkheologicheskaia progulka po moskovskim ulitsam." *Russkii arkhiv,* bk. 1 (1878): 154–72.

Martynov, I. F. "Kniga v russkoi provintsii 1760–1790-kh gg.: Zarozhdenie provintsial'noi knizhnoi torgovli." In *Kniga v Rossii do serediny XIX v.,* edited by A. A. Sidorov and S. P. Luppov. Leningrad: Nauka, 1978.

———. "Masonskie rukopisi v sobranii Biblioteki AN SSSR." In *Materialy i soobshcheniia po fondam otdela rukopisnoi i redkoi knigi BAN SSSR,* pt. 2. Leningrad: Nauka, 1978.

———. *Opisanie rukopisnogo otdela BAN SSSR.* Vol. 4, Installment 2. *Stikhotvoreniia, romansy, poemy i dramaticheskie sochineniia XVIII–per. tret' XIX v.* Leningrad: Izdatel'stvo Akademii nauk SSSR, 1980.

———. "Rannie masonskie stikhi i pesni v sobranii Biblioteki Akademii nauk SSSR (k istorii literaturno-obshchestvennoi polemiki 1760-kh gg.)." In *Russia and the World of the Eighteenth Century,* edited by R. P. Bartlett, A. G. Cross, and Karen Rasmussen. Columbus, Ohio: Slavica, 1988.

Martynov, I. F., and I. A. Shanskaia. "Otzvuki literaturno-obshchestvennoi polemiki 1750-kh godov v russkoi rukopisnoi knige. (Sbornik A. A. Rzhevskogo)." In *XVIII vek.* Vol. 11. Leningrad: Nauka, 1976.

McArthur, Gilbert H. "Freemasonry and Enlightenment in Russia: The Views of N. I. Novikov." *Canadian-American Slavic Studies* 14, no. 3 (fall 1980): 361–75.

McGrew, Roderick E. *Paul I of Russia, 1754–1801.* New York: Oxford University Press, 1992.

McIntosh, Christopher. *The Rose Cross and the Age of Reason: Eighteenth-Century Rosicrucianism in Central Europe and Its Relationship to the Enlightenment*. Leiden: E. J. Brill, 1992.

Mel'gunov, S. P., and N. P. Sidorov, eds. *Masonstvo v ego proshlom i nastoiashchem*. 2 vols. 1914–1915. Reprint, Moscow: SP "IKPA," 1991.

Miliukov, Pavel. "Educational Reforms." In *Catherine the Great: A Profile*, edited by Marc Raeff. New York: Hill and Wang, 1972.

———. *Ocherki po istorii russkoi kul'tury*. Vol. 3. Paris: Sovremennyia zapiski, 1930.

Mullan, John. *Sentiment and Sociability: The Language of Feeling in the Eighteenth Century*. New York: Oxford University Press, 1988.

Napier, J. R. and P. H. Napier. *The Natural History of Primates*. London: British Museum, 1985.

Nataf, André. *The Wordsworth Dictionary of the Occult*. Ware, Hertfordshire: Wordsworth, 1994.

Nebel, Henry M., Jr. *N. M. Karamzin: A Russian Sentimentalist*. The Hague: Mouton, 1967.

Nekrasov, Sergei. *Apostol dobra. Povestvovanie o N. I. Novikove*. Moscow: Russkii put', 1994.

Nemirovskii, A. I., and V. I. Ukolova. *Svet zvezd, ili Poslednii russkii rozenkreitser*. Moscow: Progress/Kul'tura, 1994.

Netting, Anthony Graham. "Russian Liberalism: The Years of Promise, 1842–1855." Ph. D. diss., Columbia University, 1967.

Nipperdey, Thomas. *Gesellschaft, Kultur, Theorie*. Kritische Studien zur Geschichtswissenschaft, 18. Göttingen: Vandenhoeck und Ruprecht, 1976.

Norton, Barbara T. "Russian Political Masonry and the February Revolution of 1917." *International Review of Social History* 28, pt. 2 (1983): 240–58.

Oestreich, Gerhard. *Neostoicism and the Early Modern State*. Edited by Brigitta Oestreich and H. G. Koenigsberger. Translated by David McLintock. Cambridge: Cambridge University Press, 1982.

Okenfuss, Max J. "Education and Empire: School Reform in Enlightened Russia." *Jahrbücher für Geschichte Osteuropas*, n.s., 27, Heft 1 (1979): 41–68.

———. "The Jesuit Origns of Petrine Education." In *The Eighteenth Century in Russia*, edited by J. G. Garrard. Oxford: Clarendon Press, 1973.

———. "Popular Educational Tracts in Enlightenment Russia: A Preliminary Survey." *Canadian-American Slavic Studies* 14, no. 3 (fall 1980): 307–26.

———. *The Rise and Fall of Latin Humanism in Early-Modern Russia: Pagan Authors, Ukrainians, and the Resiliency of Muscovy*. Leiden: E. J. Brill, 1995.

Orlov, Vl. *Russkie prosvetiteli 1790–1800-kh godov*. N.p.: Gosudarstvennoe izdatel'stvo khudozhestvennoi literatury, 1950.

Outram, Dorinda. *The Enlightenment*. Cambridge: Cambridge University Press, 1995.

Papmehl, K. A. "The Empress and 'Un Fanatique': A Review of the Circumstances Leading to the Government Action against Novikov in 1792." *Slavonic and East European Review* 68, no. 4 (October 1990): 665–91.

Pekarskii, P. *Dopolneniia k istorii masonstva v Rossii XVIII stoletiia*. St. Petersburg: Imperatorskaia Akademiia nauk, 1869.

Piksanov, N. K. "Masonskaia literatura." In *Istoriia russkoi literatury*. Vol. 4. Moscow-Leningrad: Izdatel'stvo Akademii nauk SSSR, 1947.

Pipes, Richard. *Russia under the Old Regime.* New York: Scribner, 1974.

Platonov, O. A. *Trenovyi venets Rossii: Istoriia masonstva, 1731–1995.* Moscow: Rodnik, 1995.

Plekhanov, G. V. *Istoriia russkoi obshchestvennoi mysli.* bk. 3. Moscow-Leningrad: Gosudarstvennoe izdatel'stvo, 1925.

Pocock, J. G. A. "Clergy and Commerce: The Conservative Enlightenment in England." In *L'età dei lumi: Studi storici sul settecento europeo in onore di Franco Venturi* 1. Naples, 1985.

———. *The Machiavellian Moment: Florentine Political Thought and the Atlantic Republican Tradition.* Princeton: Princeton University Press, 1975.

Popov, N. "Pridvornyia propovedi v tsarstvovanie Elisavety Petrovny." *Letopisi russkoi literatury i drevnosti* 2, bk. 3 (1859): 1–32.

Pozdneev, A. V. "Rannie masonskie pesni." *Scando-Slavica* 8 (1962): 26–64.

Pyliaev, M. I. *Staraia Moskva: Rasskazy iz byloi zhizni pervoprestol'noi stolitsy.* 1891. Reprint, Moscow: Moskovskii rabochii, 1990.

———. *Staryi Peterburg.* 1889. Reprint, Leningrad: Titul, 1990.

Pypin, A. N. *Russkoe masonstvo. XVIII i pervaia chetvert' XIX v.* Petrograd: OGNI, 1916.

Raeff, Marc. *Imperial Russia, 1682–1825: The Coming of Age of Modern Russia.* New York: Knopf, 1971.

———. *Origins of the Russian Intelligentsia: The Eighteenth-Century Nobility.* New York: Harcourt, Brace and World, 1966.

———. "Transfiguration and Modernization: The Paradoxes of Social Disciplining, Paedagogical Leadership, and the Enlightenment in Eighteenth-Century Russia." In *Alteuropa—Ancien Régime—Frühe Neuzeit: Probleme und Methoden der Forschung,* edited by Hans Erich Bödeker and Ernst Hinrichs. Stuttgart: F. Frommann, 1991.

———. *Understanding Imperial Russia: State and Society in the Old Regime.* Translated by Arthur Goldhammer. New York: Columbia University Press, 1984.

———. *The Well-Ordered Police State: Social and Institutional Change through Law in the Germanies and Russia, 1600–1800.* New Haven: Yale University Press, 1983.

Ransel, David L. *The Politics of Catherinian Russia: The Panin Party.* New Haven: Yale University Press, 1975.

Reinalter, Helmut. *Aufgeklärter Absolutismus und Revolution: Zur Geschichte des Jakobinertums und der frühdemokratischen Bestrebungen in der Habsburgmonarchie.* Vienna: H. Böhlau, 1980.

Reinalter, Helmut, ed. *Freimaurer und Geheimbünde im 18. Jahrhundert in Mitteleuropa.* Frankfurt a. M.: Suhrkamp, 1983.

Rétat, Pierre. "Citoyen—Sujet, Civisme." In *Handbuch der politisch-sozialer Grundbegriffe in Frankreich, 1680–1820,* edited by Rolf Reichardt and Eberhard Schmitt, Heft 9. Munich: R. Oldenbourg, 1988.

Riasanovsky, Nicholas V. *A History of Russia.* 4th ed. New York: Oxford University Press, 1984.

———. *A Parting of Ways: Government and the Educated Public in Russia, 1801–1855.* Oxford: Clarendon Press, 1976.

Robbins, Bruce, ed. *The Phantom Public Sphere.* Minneapolis: University of Minnesota Press, 1993.

Roberts, J. M. *The Mythology of the Secret Societies.* London: Secker and Warburg, 1972.

Roberts, Marie Mulvey. "Masonics, Metaphor and Misogyny: A Discourse of Marginality?" In *Languages and Jargons: Contributions to a Social History of Language,* edited by Peter Burke and Roy Porter. Cambridge: Polity Press, 1995.

Roche, Daniel. "Literarische und geheime Gesellschaftsbildung im vorrevolutionären Frankreich: Akademien und Logen. In *Lesegesellschaften und bürgerliche Emanzipation: Ein europäischer Vergleich,* edited by Otto Dann. Munich: C. H. Beck, 1981.

Rogger, Hans. *National Consciousness in Eighteenth-Century Russia.* Cambridge, Mass.: Harvard University Press, 1960.

Romanovich-Slavatinskii, A. *Dvorianstvo v Rossii ot nachala XVIII veka do otmeny krepostnogo prava.* 2nd ed. Kiev: Universitet Sv. Vladimira, 1912.

Roosevelt, Priscilla. *Life on the Russian Country Estate: A Social and Cultural History.* New Haven: Yale University Press, 1995.

Rozman, Gilbert. *Urban Networks in Russia, 1750–1800, and Premodern Periodization.* Princeton: Princeton University Press, 1976.

Ryu, In-Ho L. "Freemasonry under Catherine the Great: A Reinterpretation." Ph.D. diss., Harvard University, 1967.

———. "Moscow Freemasons and the Rosicrucian Order. A Study in Organization and Control." In *The Eighteenth Century in Russia,* edited by J. G. Garrard. Oxford: Clarendon Press, 1973.

Sanders, J. Thomas. "The Third Opponent: Dissertation Defenses and the Public Profile of Academic Life in Late Imperial Russia." *Jahrbücher für Geschichte Osteuropas,* n.s., 41, Heft 2 (1993): 242–66.

Schneewind, J. B. Introduction to *Moral Philosophy from Montaigne to Kant: An Anthology.* 2 vols. Cambridge: Cambridge University Press, 1990.

Seligman, Adam B. *The Idea of Civil Society.* New York: Free Press, 1992.

Semeka, A. V. "Russkie rozenkreitsery i sochineniia imperatritsy Ekateriny II protiv masonstva." *Zhurnal Ministerstva narodnago prosveshcheniia,* pt. 39, no. 2 (1902): 343–400.

Semennikov, V. P. "Literaturno-obshchestvennyi krug Radishcheva." In *Radishchev: Materialy i issledovaniia,* edited by A. S. Orlov. Moscow-Leningrad: Izdatel'stvo Akademii nauk SSSR, 1936.

Semenova, L. N. *Ocherki istorii byta i kul'turnoi zhizni Rossii (pervaia polovina XVIII v.).* Leningrad: Nauka, 1982.

Serkov, A. I. "Pridet li k nam 'Velikii vostok'?" *Rodina* 11 (1993): 85–87.

———. "Rossiiskoe masonstvo: Entsiklopedicheskii slovar'. Vol. 1. Masony v Rossii (1731–1799)." Unpublished manuscript. Moscow, n.d.

Shamrai, D. D. "Ob izdateliakh pervogo chastnogo russkogo zhurnala (Po materialam arkhiva kadetskogo korpusa)." In *XVIII vek.* Moscow-Leningrad: Izdatel'stvo Akademii nauk SSSR, 1935.

Shevyrev, S. P. *Istoriia Imperatorskago Moskovskogo Universiteta.* Moscow: Universitetskaia Tipografiia, 1855.

Shmidt, S. O. "Obshchestvennoe samosoznanie Noblesse russe v XVI–pervoi treti XIX v." *Cahiers du monde russe et soviétique* 34, nos. 1–2 (January–June 1993): 11–32.

Shmurlo, E. *Mitropolit Evgenii kak uchenyi. Rannie gody zhizni, 1767–1804*. St. Petersburg: Tipografiia V. S. Balasheva, 1888.

Shtrange, M. M. *Russkoe obshchestvo i frantsuzskaia revoliutsiia, 1789–1794 gg.* Moscow: Akademiia nauk, 1956.

Simmel, Georg. *The Sociology of Georg Simmel*. Edited and translated by Kurt H. Wolff. Glencoe, Ill.: Free Press, 1950.

Skabichevskii, A. M. *Ocherki istorii russkoi tsenzury (1700–1863 g.)*. St. Petersburg: Obshchestvennaia pol'za, 1892.

Slovar' russkikh pisatelei XVIII veka. Vol. 1. Edited by A. M. Panchenko. Leningrad: Nauka, 1988.

Smirnov, N. A. "Zapadnoe vliianie na russkii iazyk v petrovskuiu epokhu." *Sbornik otdeleniia russkago iazyka i slovesnosti Imperatorskoi Akademii nauk* 88, no. 2 (1910): 1–399.

Smith, Nathan. "Political Freemasonry in Russia, 1906–1918: A Discussion of the Sources." *Russian Review* 44, no. 2 (April 1985): 157–73.

Sokolovskaia, Tira. "Ioannov den'—masonskii prazdnik." *More* 23–24 (1906): 846–63.

———. "Masonskie kovry (Stranichka iz istorii masonskoi simvoliki)." *More* 13–14 (1907): 417–34.

———. "Masonstvo v teorii i v zhizni." Pts. 1 and 2. *More* 11–12 (1906): 392–97; 13–14 (1906): 479–91.

———. "O masonstve v prezhnem russkom flote." *More* 8 (1907): 216–53.

———. *Russkoe masonstvo i ego znachenie v istorii obshchestvennogo dvizheniia (XVIII i pervaia chetvert' XIX stoletiia)*. St. Petersburg: n.p., [1907].

———. "Traurnaia lozha u masonov (K istorii masonstva v Rossii)." *More* 6 (1906): 205–11.

Solov'ev, O. F. *Russkoe masonstvo, 1730–1917*. Moscow: Izdatel'stvo MGOU, 1993.

Staniukovich, T. V. *Kunstkamera peterburgskoi Akademii nauk*. Moscow-Leningrad: Akademiia nauk SSSR, 1953.

Starikova, L. *Teatral'naia zhizn' starinnoi Moskvy: Epokha, byt, nravy*. Moscow: Iskusstvo, 1988.

Starobinski, Jean. *Blessings in Disguise; or, The Morality of Evil*. Translated by Arthur Goldhammer. Cambridge, Mass.: Harvard University Press, 1993.

Steiner, Gerhard. *Freimaurer und Rosenkreuzer: Georg Forsters Weg durch Geheimbünde*. Berlin: Akademie Verlag, 1985.

Stepanskii, A. D. *Istoriia obshchestvennykh organizatsii dorevoliutsionnoi Rossii*. Moscow: MGIAI, 1979.

Stevenson, David. *The Origins of Freemasonry: Scotland's Century, 1590–1710*. Cambridge: Cambridge University Press, 1988.

Stites, Richard. "Old Smolensk: Two Sketches and Notes for Research." Paper presented at the workshop "Visions, Institutions, and Experiences of Imperial Russia," Washington, D.C., September 1993.

Stoletie S. Peterburgskago angliiskago sobraniia, 1770–1870. St. Petersburg: V. I. Golovin, 1870.

Sushkov, N. V. *Vospominaniia o Moskovskom universitetskom blagorodnom pansione*. Moscow, 1848.

Svodnyi katalog russkoi knigi grazhdanskoi pechati XVIII veka, 1725–1800. 5 vols., 1 supplement. Moscow: Kniga, 1962–1975.

Tarasov, E. "K istorii russkago obshchestva vtoroi poloviny XVIII stoletiia: Mason I. P. Turgenev." *Zhurnal Ministerstva narodnago prosveshcheniia*, n.s., 51, no. 6, section 3 (1914): 129–75.

Taylor, Charles. *Philosophical Arguments*. Cambridge, Mass.: Harvard University Press, 1995.

———. *Sources of the Self: The Making of the Modern Identity*. Cambridge, Mass.: Harvard University Press, 1989.

Taylor, Katrina V. H. *Russian Art at Hillwood*. Washington, D.C.: Hillwood Museum, 1988.

Todd, William Mills, III. *Fiction and Society in the Age of Pushkin: Ideology, Institutions, and Narrative*. Cambridge, Mass.: Harvard University Press, 1986.

Tukalevskii, Vl. "Iz istorii filosofskikh napravlenii v russkom obshchestve XVIII veka." *Zhurnal Ministerstva narodnago prosveshcheniia*, n.s., 33, no. 5 (1911): 1–69.

Turner, Victor. *The Ritual Process: Structure and Anti-Structure*. Ithaca, N.Y.: Cornell University Press, 1969.

Utkina, N. F., et al. *Russkaia mysl' v vek prosveshcheniia*. Moscow: Nauka, 1991.

Varnecke, B. V. *History of the Russian Theatre*. New York: MacMillan, 1951.

Vernadskii, G. V. *Russkoe masonstvo v tsarstvovanie Ekateriny II*. Petrograd: Istoriko-filologicheskii fakul'tet Petrogradskago universiteta, 1917.

Vsevolodskii-Gerngross, V. *Istoriia russkogo dramaticheskogo teatra*. Vol. 1, *Ot istokov do kontsa XVIII veka*. Moscow: Iskusstvo, 1977.

———. *Russkii teatr ot istokov do serediny XVIII veka*. Moscow: Izdatel'stvo Akademii nauk SSSR, 1957.

Weisberger, R. William. *Speculative Freemasonry and the Enlightenment: A Study of the Craft in London, Paris, Prague, and Vienna*. East European Monographs, 367. New York: Columbia University Press, 1993.

White, Hayden. "The Forms of Wildness: Archeology of an Idea." In *The Wild Man Within: An Image in Western Thought from the Renaissance to Romanticism*, edited by Edward Dudley and Maximillian E. Novak. Pittsburgh: University of Pittsburgh Press, 1972.

Wilding, Peter. *Adventurers in the Eighteenth Century*. New York: G. P. Putnam's Sons, [1937].

Wirtschafter, Elise Kimerling. *Social Identity in Imperial Russia*. DeKalb, Ill.: Northern Illinois University Press, 1997.

———. *Structures of Society: Imperial Russia's "People of Various Ranks."* DeKalb, Ill.: Northern Illinois University Press, 1994.

Wolff, Larry. *Inventing Eastern Europe: The Map of Civilization on the Mind of the Enlightenment*. Stanford: Stanford University Press, 1994.

Wuthnow, Robert. *Communities of Discourse: Ideology and Social Structure in the Reformation, the Enlightenment, and European Socialism*. Cambridge, Mass.: Harvard University Press, 1989.

Yates, Frances A. *Giordano Bruno and the Hermetic Tradition*. Chicago: University of Chicago Press, 1964.

———. *The Rosicrucian Enlightenment*. London: Routledge and Kegan Paul, 1972.

Zabelin, I. "Iz khroniki obshchestvennoi zhizni v Moskve v XVIII stoletii." In *Sbornik obshchestva liubitelei rossiiskoi slovesnosti na 1891 god*. Moscow: I. N. Kusherev, 1891.

————. "Khronika obshchestvennoi zhizni v Moskve s poloviny XVIII stoletiia." In *Opyty izucheniia russkikh drevnostei i istorii*, pt. 2. Moscow: Tipografiia Gracheva, 1873.

Zaitseva, A. A. "'Kabinety dlia chteniia' v Sankt-Peterburge kontsa XVIII–nachala XIX veka." In *Russkie biblioteki i chastnye knizhnye sobraniia XVI–XIX vekov*, edited by D. V. Ter-Avanesian. Leningrad: BAN SSSR, 1979.

Zapadov, A. V. "Iv. Novikov, Komarov, Kurganov." In *Istoriia russkoi literatury*. Vol. 4. Moscow-Leningrad: Izdatel'stvo Akademii nauk SSSR, 1947.

————. "Tvorchestva Kheraskova." In *Izbrannye proizvedeniia*, by M. M. Kheraskov. Leningrad: Sovetskii pisatel', 1961.

Zolotov, Oleg. "Russian Masons True to Their Homeland." *Moscow News*, no. 8 (25 February–3 March 1994): 15.

Zorin, Andrei. "K predystorii odnoi global'noi kontseptsii (Oda V. P. Petrova 'Na zakliuchenie s Ottomanskoiu Portoiu mira' i evropeiskaia politika 1770-kh godov)." *Novoe literaturnoe obozrenie* 23 (1997): 56–77.

Zotov, V. "Kaliostro, ego zhizn' i prebyvanie v Rossii." *Russkaia starina* 12 (January 1875): 50–83.

Index